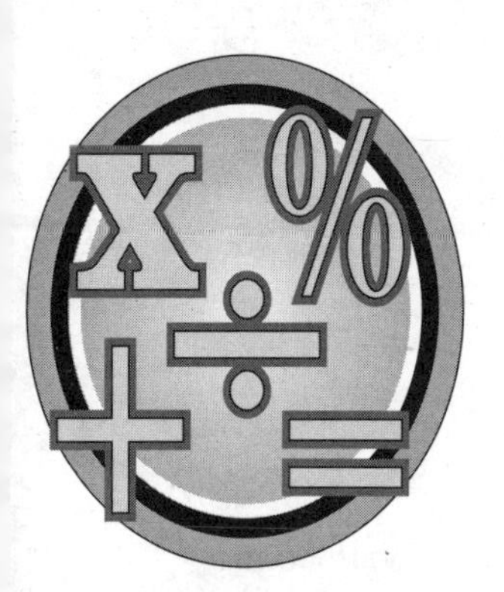

# EVERYDAY MATH MADE E-Z

Pete Reinert

**MADE E-Z PRODUCTS, Inc.**
Deerfield Beach, Florida / www.MadeE-Z.com

Everyday Math Made E-Z™

900 East Wayzata Boulevard, Wayzata, MN 55391-1836
Tel. 612-476-9200
Fax 612-475-2373
http://www.LearningStrategies.com


1 2 3 4 5 6 7 8 9 10 CPC R 10 9 8 7 6 5 4 3 2

This publication is designed to provide accurate and authoritative information in regard to subject matter covered. It is sold with the understanding that neither the publisher nor author is engaged in rendering legal, accounting, or other professional services. If legal advice or other expert assistance is required, the services of a competent professional should be sought. From: *A Declaration of Principles jointly adopted by a Committee of the American Bar Association and a Committee of Publishers.*

Everyday Math Made E-Z™
Pete Reinert

## Limited warranty and disclaimer

This self-help product is intended to be used by the consumer for his/her own benefit. It may not be reproduced in whole or in part, resold or used for commercial purposes without written permission from the publisher. In addition to copyright violations, the unauthorized reproduction and use of this product to benefit a second party may be considered the unauthorized practice of law.

This product is designed to provide authoritative and accurate information in regard to the subject matter covered. However, the accuracy of the information is not guaranteed, as laws and regulations may change or be subject to differing interpretations. Consequently, you may be responsible for following alternative procedures, or using material or forms different from those supplied with this product. It is strongly advised that you examine the laws of your state before acting upon any of the material contained in this product.

As with any matter, common sense should determine whether you need the assistance of an attorney. We urge you to consult with an attorney, qualified estate planner, or tax professional, or to seek any other relevant expert advice whenever substantial sums of money are involved, you doubt the suitability of the product you have purchased, or if there is anything about the product that you do not understand including its adequacy to protect you. Even if you are completely satisfied with this product, we encourage you to have your attorney review it.

Neither the author, publisher, distributor nor retailer are engaged in rendering legal, accounting or other professional services. Accordingly, the publisher, author, distributor and retailer shall have neither liability nor responsibility to any party for any loss or damage caused or alleged to be caused by the use of this product.

## Copyright Notice

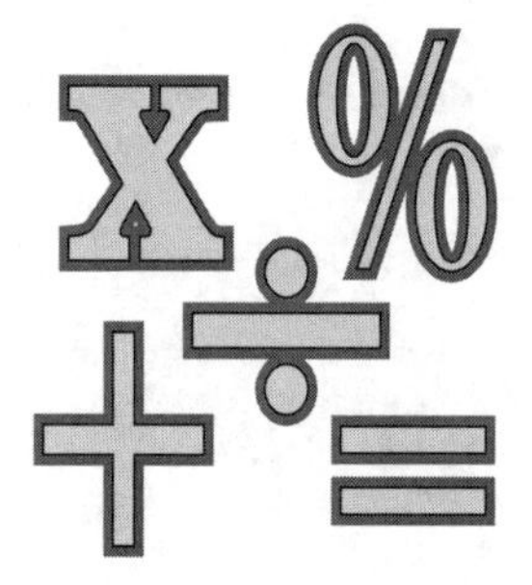

# Table of contents

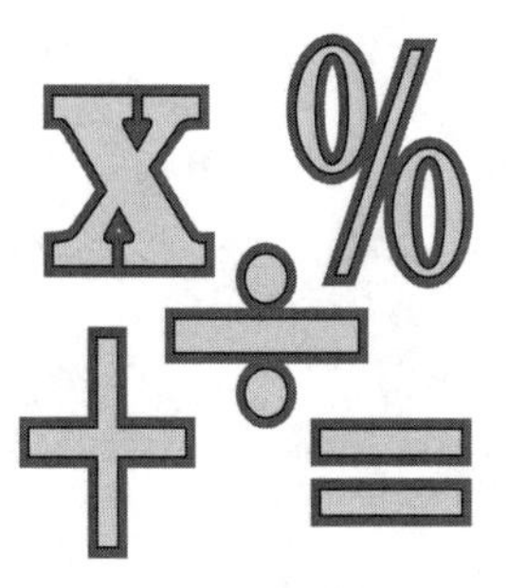

# Introduction to Everyday Math Made E-Z™

How do you calculate a 15% tip for a $35 meal? Is a 20-oz. bottle of shampoo at $1.98 a better value than a 35-oz. bottle at $3.95? Is the sale you see featured in the newspaper really a bargain or just a public relations gimmick?

Common sense is not enough to answer these questions—you need good, basic math skills to solve everyday problems like these. But what if you don't have those skills? What if, in school, you had trouble understanding and approaching math work? What if you suffer from "math block"?

With *Everyday Math Made E-Z*, you don't have to be a math whiz to understand and use basic math skills everyone needs. In this guide you'll find all the tips and tricks of the math masters using only the basics of math: addition, subtraction, multiplication, and division. You'll also learn tips on how to approach math without frustration or fear.

A workbook of sample problems, skill drills, and word problems is included with plenty of space for working out the problems—and all the answers are here, too. Plus, you'll find a Resources section with where to turn for more information.

No matter how difficult learning math was in the past, now you can master the skills you need to become a human calculator—with *Everyday Math Made E-Z*!

# E-Z math secrets

# 1

# Chapter 1

## E-Z math secrets

## Secrets to E-Z math

Why do you need to learn math? Good question. After all, your bank takes care of the money you have in your checking account and the interest due you from savings. Your credit card company keeps track of how much you charge and the interest owed them. Unless your income is from just a few simple sources, you probably have an accountant figure out your taxes. If you wanted to, you could get through life perfectly well without math. But it would cost you.

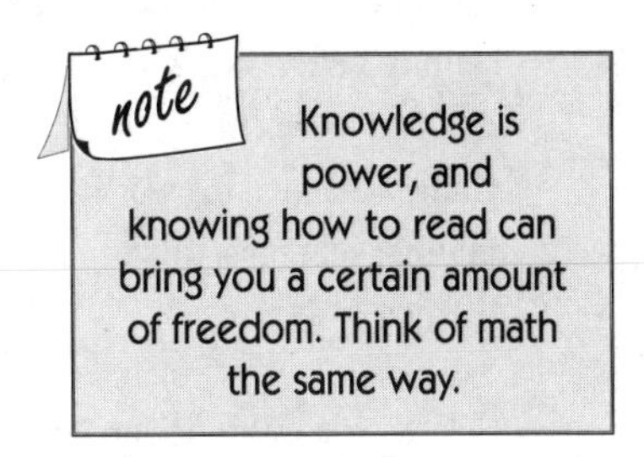


Imagine going through life without knowing how to read. Many people do, and some make out pretty well. Many do not, though. Being illiterate in today's world is a significant disadvantage because it restricts the information one is able to access, making informed decisions more difficult. When I returned to college after an absence of five years or so, I hadn't done much math in almost a decade, and my college algebra class was driving me nuts! When I mentioned my problem to my father, he said: "What you are doing right now is essentially learning a new language. It takes awhile to get used to speaking and thinking in that language."

Or something like that. I really was not listening carefully because my interaction with math had reduced me to a basket case. At the time, I did not find those words very useful, but they did relieve the stress of my situation somewhat. Instead of letting my eyes get tired and watery when I faced an opponent like:

$$f(x) = x^3 + 2x^2 + 4x;\ f(4)$$

I would take a deep breath and start figuring it out, step by step. And after working a number of problems, I got to the point where it was instinctive, or implicit. In the same way that my brain developed a program to tie my shoe when I was five, my brain had set up a system to figure out math—I had learned the language. The answer, by the way, is 112.

But we have gotten away from why it is a good idea to learn math in the first place. Unless you are still in school, it may not seem like math is all that necessary.

Do you want the credit card which offers 18 percent interest and no annual fee, or the one that charges 12 percent with a $50 annual fee? Instinctively, the 12 percent one sounds better, but how much better, exactly? How many hours are you going to have to work to pay the interest charges, given an average balance of $750?

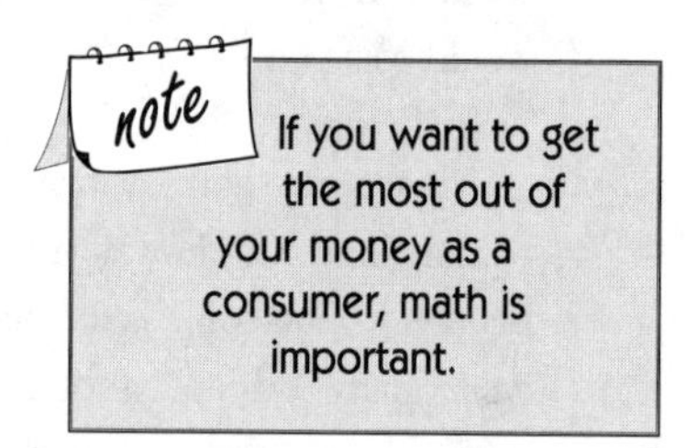


If you want to understand the real issue behind the murky veil of statistics, math is important. President Porkbarrel is leading his rival, Senator Hotair, by 3 percent in the polls. But the survey has a margin of error of 5%. Does this mean Hotair could be winning? How will this affect your business? Your investments?

If you need to plan what a trip will cost you or your company, math is important. Imagine being stuck in a strange city without the money for a hotel room, simply because your flight, taxi ride, meals and convention registration cost more than you thought they would.

If you would like to know how many miles your car gets per gallon, math is important. Suddenly it seems like you're filling up the car a lot more frequently than usual, which might indicate a fuel leak or another mechanical problem. If you had kept a mental note of how many miles to the gallon you had been getting, you might now have more evidence to support your suspicion.

It is probably starting to become apparent to you that math is all around us. When I was in high school, trying to fit three study halls into my class schedule, my father said "I recommend that you take all the math and all the hard science you can get your hands on."

At the time, I was planning on going to college for art, so I ignored him, as teenagers will do. But time has a way of changing one's point of view, and when I started college a few years later, it was as a physics minor.

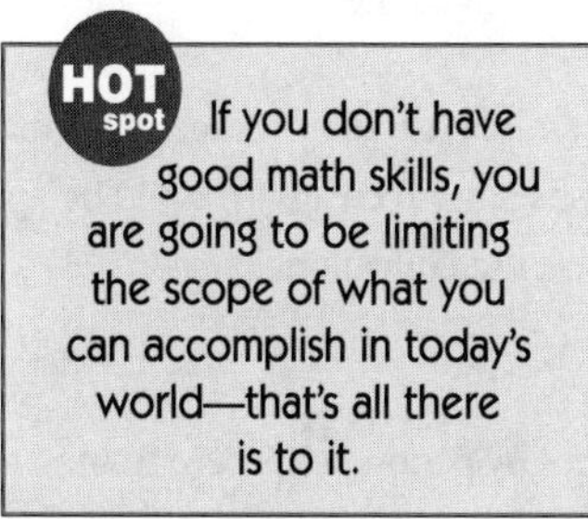


At the risk of sounding paranoid, I will offer an opinion: Many people in today's world do not want you to have good math skills. Companies learned a long time ago that they sold more big screen TVs if they offered payment plans of $50 per week (for only 40 weeks, making a total of $2,000), rather than trying to sell them for $1,000, cash on delivery. That's an awful big difference in price just for the convenience of making weekly payments.

Perhaps, after frightening you with what may happen if you don't know math, I should add that math can be a joy. There is nothing so elegant as a properly worked math problem, and when you finish working a particularly involved equation and you check your answer with the one in the back of the book and find that your answer is correct, you get a little "high" from the experience.

Math never came easy to me when I was a kid. I got Bs and Cs which were not considered good in my family, and I worried a lot to get them. I think part of the problem was having to live up to that weird standard that kids' peers lay upon them—you don't want to appear to be too smart, or you're a "brain" (i.e. a social leper).

The standard parents lay on kids is just as bad—"Get passing grades, or it is (fill in favorite punishment here) for you!" Thus, many kids are stuck at "C level" and end up getting too neurotic to learn much of anything. It is a crazy system, but it endures.

It was not until I got to college that I started to suspect that my lack of knowledge was not all my fault. By that time, my father had spent several years learning accelerative learning techniques and using them in his physics class, so I walked into the educational factory with a little more knowledge about what learning was supposed to be and the circumstances under which it worked best than they probably would've liked me to have.

note

I knew that learning worked best when it was fun, interesting, maybe set to music or rhythm, and when a person's natural creativity could be applied—in short, the same way kids learn before the educational system gets a hold of them.

What I got, generally, were lectures by people who were trained to know their fields, but not to teach—any grade school teacher has had more training on how to convey information to students than almost any Ph.D. I also got textbooks written by people who must have been extras in *Night of the Living Dead.*

Don't get me wrong—there are many fine teachers out there who fulfill the duties of a "learning coach." But there are a lot of crummy ones, too. I think that at least part of the reason that math and I did not get along for so many years was because of rotten teaching. The math teacher I had in third grade used to dig her claws into my shoulder whenever I answered a question wrong. I defy anyone to learn effectively under such circumstances.

Quick—what letter comes three letters after "M" in the alphabet?

Notice how you found the answer—you probably sang *The Alphabet Song* in your head, or hummed it, until you got to three letters past "M." When you learned the alphabet, you probably learned it by singing *The Alphabet Song.* It was fun, easy, and it brought a sense of accomplishment to your younger self. Can you imagine a parent who would look down at a two- or three-year-old child and say, "Here is a list of all the letters in the alphabet. I expect you to memorize them, in order, by Friday. There will be a test, and should you fail it, you will have to be two years old for another year."

That is crazy! No one teaches their kids things in that manner. But that's the way the system has been teaching kids for at least 100 years. Shut up, sit still, get permission to ask a question, do not question what the teacher is saying, and no singing or whistling except in music class. All of those commands could just as well be carved into the blackboard in nearly every class in the country. If I was trying to ensure that a group of people were going to be good drones to stand at machines in my factory, that is exactly the way I'd set up a class, but not if I wanted to stay in business very long. And certainly not if I was trying to teach future architects, musicians, electrical engineers, advertising reps or psychologists.

Here's another interesting thing I learned in college: Every year schoolteachers will get together and tell each other who the "good" and "bad" students are. It does not take a psychologist to guess that "bad" students get lower grades than "good" students, even when the work they turn in is exactly the same! Sometimes you do not even have to set the standard yourself—it is often enough just to have an older brother or sister who is a "bad student" in order to reap the benefits for yourself. Nice system!

My point is that you can be a math whiz even if all through your educational career you were told that you were a "bad student." Sometimes parents will help out on this one:

"My Roger is not very bright, but he's popular!"

Ever hear Mom or Dad say this while you were standing right in front of them? When a kid repeatedly hears statements like that from respected adults he starts to believe them and thus starts to act like them. But it is never too late to change.

> **HOT spot** Even if you have been told that you were stupid and terrible in math, I promise that you can easily learn math. And you may just enjoy it.

You may have tried other rapid-fire math systems in the past that have all sorts of complicated techniques you need to memorize in order to make math easier. This really isn't that sort of book. There are some tricks to math and we'll give you some. But having to memorize 75 different rules for multiplication of two different numbers, depending on what they are, really does not sound like any definition of the word "easy" I have heard lately. Sometimes the old-school system is best, but I will make it easier when I can.

Because I do not know at what level you are currently using math, we will start with the absolute basics, play around with a little probability and statistics, and dabble with algebra, which is really as far as most people (that is, non-rocket scientists) need to go on the path to total math enlightenment. We will then take a look at applying this math to your daily life.

We will skip most of the technical terms, when we can, like "divisor" and "quotient." I do not think anyone but math teachers ever uses these words, and while it is nice to know that there's a name for the number that goes on the bottom of a fraction, it is really not necessary to know in order to get the problem right.

Speaking of words, I cannot imagine why anybody started calling mathematical equations "problems." Cancer is a problem. An IRS audit is a problem. But math? Math is just a puzzle, something to be worked out. In adhering to the low-stress attitude for this course, we will call our work not "problems" but "puzzles." You may think it is just a matter of semantics—that what we call something doesn't make any difference—but consider how emotionally laden the word "test" is compared to the word "quiz." Each is a method of testing knowledge, but many people sweat a lot more when they hear the first word than the second.

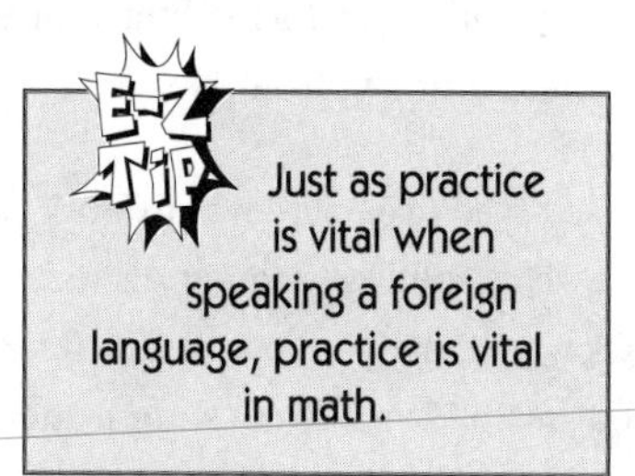


While reading the sections will provide you with a certain understanding of what is going on, I cannot guarantee that merely reading will give you the math skills you need.

Practicing in the workbook may take you half an hour per night, or an hour, or two hours. But once you have gotten those skills down, you've got them for life. So sharpen up your pencil and fetch a fresh eraser, and we will get busy.

## Before beginning

Before we proceed to the rest of the book, I feel it is important to mention some words about how to learn in the most efficient manner:

Make sure to "set the mood," if you can, before beginning to work on math. Setting the mood means closing the door on distractions, turning off the ringer on the phone, and, if you have the luxury, putting on some classical music.

Listening to classical music, especially "baroque" style, has been clinically proven to raise your I.Q. temporarily. Not only that, but it can be very relaxing. We recommend a state of "alert relaxation" whenever someone is trying to learn a new thing because it makes learning easier. Here are some other tips to boost your performance:

☆ Drink plenty of water. Today, many people start with coffee in the morning, drink soda at lunch, and perhaps have a beer when they get home from work. All three of these drinks act as diuretics—that is, they tend to dehydrate you. Your brain, however, needs plenty of water in order for the neurotransmitters (the chemicals in your brain) to work properly. In addition, although caffeine turbocharges your brain, the "high" you get from coffee or soda is just a temporary one, and when it wears off you will probably feel "drained" until you drink some more. When I had to take a college final, I would generally drink a couple of cups of coffee, then a bunch of water. For everyday math work consider drinking fruit juice or water.

E-Z TIP

☆ Set a clear purpose for yourself when you sit down to work. Sitting down with a vague intention is a good way to get distracted. Tell yourself something like, "Okay, self, my purpose at this time is to learn as much as I can about math." It really does not matter how you phrase it, but setting a definite purpose (and telling yourself what it is) does seem to be important to success. It is really not within the scope of this book to explain why setting a definite purpose works, but if you plan to trust me to help you with math, please trust me on this aspect also. This tip, by the way, works well in any aspect of life. When I sit down to write, my stated purpose is to write the best book I possibly can (at least until the editors get their hands on it).

☆ Get plenty of sleep. Americans, at least, generally get less sleep than they need to get, which often makes them dopey and dependent on caffeine. I know that if I get less than 7 hours of sleep on any particular night, I will not be worth much the next day. You know what amount of sleep is right for you—see that you get it.

HOT spot

If you need a break, take one. If you cannot concentrate, you are really not learning anything, so it is better to do something else for a few minutes.

☆ If the words on the page are starting to blur together, or what I am saying does not make sense, feel free to take a break. Few people can maintain a state of utter concentration for an hour or so, and I am not one of them.

☆ Watch your mouth. Just as the "hidden" part of your brain (your subconscious mind, actually) listens when you set your purpose for a session, it listens to you during the rest of the day, too. Saying things like, "I find math difficult," or "Math and I don't get along," will hurt you in the long run, because your subconscious will believe you and try to bring you what it thinks you want—to have difficulty with math. As with declaring your definite purpose, this advice applies to the rest of your life, too—not just math. Skeptical? You have nothing to lose by giving it a try.

> **E-Z TIP**
> If you find yourself discussing past problems with math, say something like, "I used to have trouble with math, but I'm getting much better."

☆ When you are not working with this book, when you are at work or driving or at the dentist, feel free to think about math and play with it a little. The other day, I was sitting around, supposedly working on a magazine article, but really thinking about a heat-exchanger that a company was selling, and I found that I couldn't concentrate on writing my article until I found out what the most efficient heat-exchange ventilation system would look like.

The point is, thinking about things like that is an excellent way to make math your friend. When I work a puzzle in my head, I visualize a chalkboard and "see" the numbers set up in my mind's eye. When I was "working" on the heat exchange ventilation system, I "saw" the cross-section of the two tubes, including notation that told me how wide the pipes were.

☆ Visualizing, or "seeing" problems in your mind's eye is an excellent way to incorporate your whole brain into figuring out a math problem. You have probably heard about the "left brain, right brain" split. That is, the left side of the brain processes all the logic-oriented stuff and the right side handles the creative, artsy stuff. When you visualize or draw a picture of a math problem, you are setting both sides to work, instead of just one.

Albert Einstein used to do a lot of visualization, strolling around his neighborhood and pretending he was a photon shooting through space. Some sources say that it was the visualization that made him so brilliant. I cannot say whether visualization caused his brilliance, but it seems that an awful lot of brilliant people visualize, in one way or another. If you would like to be excellent in other areas besides math, I'd suggest you pick up the habit.

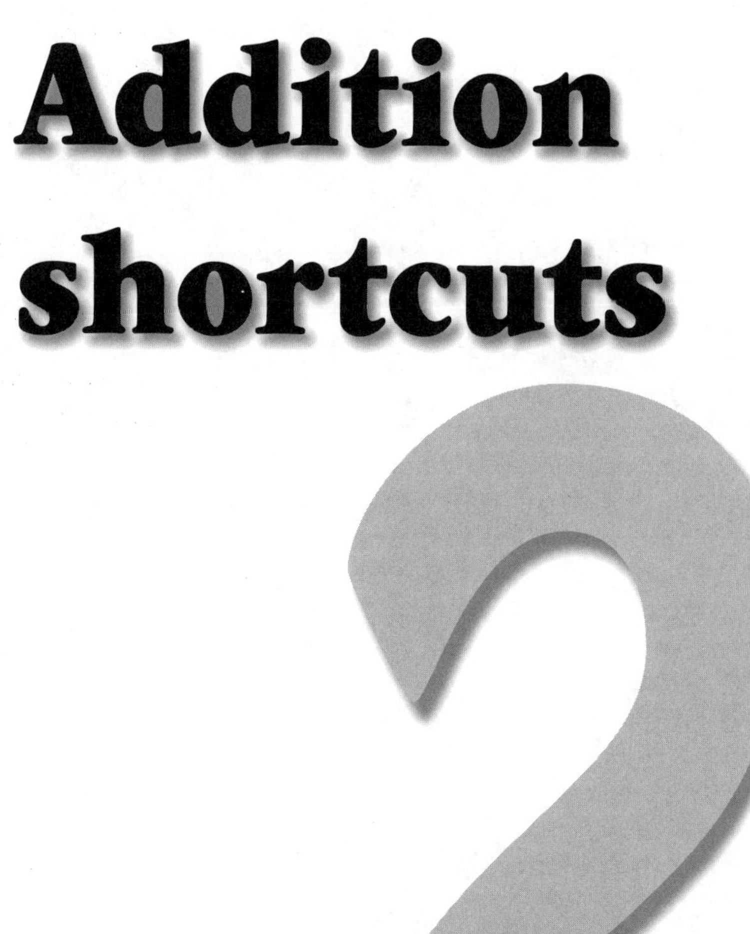
Addition
shortcuts
2

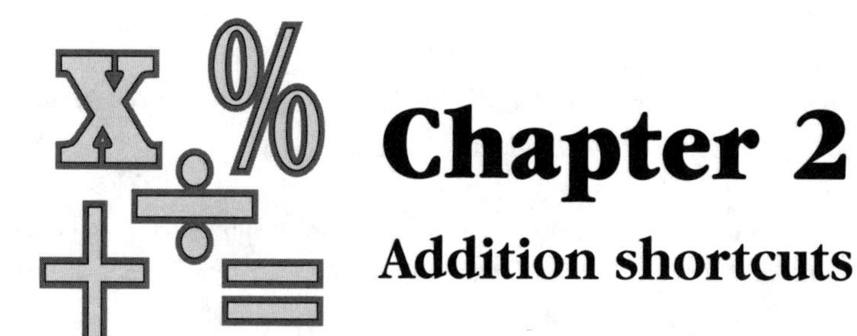

# Chapter 2
## Addition shortcuts

### *The Fudge Factor*

In the next few chapters of this book, we'll be looking at techniques for being able to do math in your head, possibly by just looking at figures on a piece of paper. When you're in the supermarket, on the highway, on a business trip, answering questions at a presentation or speech that you're giving, in the shower, on the golf course, or mowing the lawn, a calculator is just not likely to be around. It is important not to put yourself at a disadvantage just because a machine does not happen to be around.

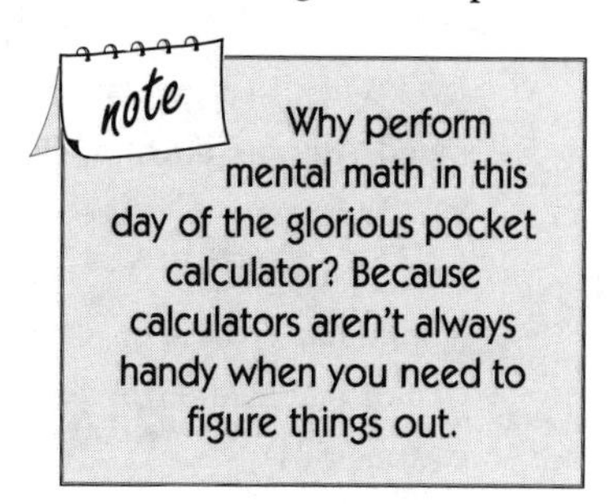


In addition, it is impossible, on most calculators, to make sure that you entered all of the digits correctly, short of punching in all the numbers again. And what if you hit the wrong number, or a button you hit did not register? In these instances, it will be faster to do a problem in your head or by looking at it on the page—and I will show you how you can check your work.

When I got my first calculator, back in eighth grade, my father told me that he always warned his students against trusting their calculators. In today's world, it seems that everyone does, but I believe it is important to know how to do the work on your own, without relying on a crutch.

One of the easiest ways to add, divide, multiply or subtract, if possible, is to round numbers off. That is, to treat a given number, say 422, as if it was really 400 or 425. Rounded numbers make your calculations speed by because it is a lot easier to count by 10s or 100s, or even by 30s, than it is to just plain add numbers.

There is nothing dishonest about it; even rocket scientists round numbers whenever they can. Depending on the degree of accuracy necessary, you can often get a pretty good figure on a stack of numbers without much effort. Just remember that when multiplying and dividing "fudged" numbers, your margin of error increases by a lot more than when adding and subtracting them.

When I balance my checkbook, I round all the checks I write up, and round deposits down to the nearest dollar. A check to the gas station for $17.32 becomes $18.00, and a deposit of $773.22 changes to $773. Not only does rounding to the nearest dollar make it easier to figure out my (approximate) balance whenever I sit down to do it, but it decreases the chance of a math error leading to an insufficient funds charge (which could cost me $24).

This system also has the pleasant side effect of building up a hidden reserve in my checkbook in case I ever forget to write down a withdrawal from a cash machine. Sloppy bookkeeping, you say? Not necessarily. It all depends on what the situation calls for.

In the case of my checkbook, I do not need to know my balance to the exact penny. All I need to know is whether I have money or not, and if so, roughly how much. I trust my bank to keep track of the pennies, and I am pretty sure I will get them back when I close my account.

When I generate a page of figures for my accountant to calculate my taxes (not even people who are super at math can always please the IRS), that's another matter. Knowing how picky the IRS can be, I provide all figures down to the last cent.

Rounding in the checkbook balancing system works differently than in most other equations, but it is such a slick little system, I just had to share it. The question you need to keep in mind is, "How accurate do I really need to be?"

If you are going to be driving to Pittsburgh from wherever you are for fun (say you're 366.25 miles away), it may serve your purposes well enough to say that you will be driving 350, 360, or 375 miles. If you are driving there for business, however, and the company is reimbursing you for mileage at $.29 per mile, it is obviously in your best interest to collect every penny you are due by logging the trip as 366.20 or 366.30 miles, depending on what the odometer says.

Generally, the mathematical standard for rounding is that 1, 2, 3 and 4 get rounded down to 0; and 5, 6, 7, 8 and 9 get rounded up to 10. You could round the last number, or you could treat 63 as 60, and round that up to 100. Again, it all depends on the accuracy you think you need. Thus, if you decided to spend your Christmas vacation driving all over the country to visit relatives, you might be looking at the following trips (in miles):

| | | | |
|---|---|---|---|
| Your house to Grandma's | 123 | rounds to: | 120 |
| Grandma's to Aunt Heather's: | 76 | rounds to: | 80 |
| Aunt Heather's to Uncle Bill's: | 235 | rounds to: | 240 |
| Uncle Bill's to Mom & Dad's: | 160 | rounds to: | 160 |
| Mom & Dad's to Home | +63 | rounds to: | 60 |
| Total: | 657 | | 660 |

Pretty close, huh? It won't always come out this close to the correct amount, but it is quicker than hauling out the calculator, and, of course, you could perform the math in your head while you are driving from place to place. Of course, if you really want to get sympathy from your friends, you could round the 660-mile figure again and tell them that you just got done driving 700 miles.

## *Making tens*

What if you had to add up a bunch of single-digit numbers in a hurry? You could plod through like they taught us in grade school, starting at the top of the column and keeping mental track of where you are, but it is a lot easier to Make Tens. Simply go up and down the column and cancel out any numbers you can add together to make a ten.

It is a lot easier to keep track of "10, 20, 30" than "27, 32, 36" You can keep track of the tens on your fingers, if you want to; it is perfectly acceptable.

Look at the following columns. Say you were faced with the problem on the left. Your first move might look like the column on the right.

$$\begin{array}{r} 7 \\ 4 \\ 3 \\ 8 \\ 9 \\ 4 \\ 2 \\ 5 \\ \underline{+\,7} \end{array} \qquad\qquad \begin{array}{r} \underline{7} \\ 4 \\ \underline{3} \\ \underline{8} \\ 9 \\ 4 \\ \underline{2} \\ 5 \\ \underline{+\,7} \end{array}$$

The numbers 7 and 3 in the right column have been underlined to indicate that they're part of a ten. Just keep counting by tens in your head, then add any leftover numbers to the result. In the above example, we can make one more ten (either 8 and 2 or 4, 4 and 2) and then we're stuck with a bunch of ugly numbers left over. Great—now what?

No problem. Whenever you get stuck adding two numbers together which total more than ten, like 8 and 9, just take the "1" from the tens column and add it to your total while mentally treating the remainder as just another number. Confused? Let's look at it on paper:

8 + 9 = 17

No mystery there. But what we're also saying when we say "8 + 9 = 17" is:

8 + 9 = 10 + 7

Therefore, we can take the 10 of the "10 + 7" and add it to the mental stack in our heads while keeping the 7 around to add to the next number, thus: 7 + 5 (the next number in the column). Got it? Good.

Just for fun, let's see what the actual answer was, and what we would have gotten had we rounded:

| Actual | Rounded |
|---|---|
| 7 | 10 |
| 4 | 0 |
| 3 | 0 |
| 8 | 10 |
| 9 | 10 |
| 4 | 0 |
| 2 | 0 |
| 5 | 10 |
| + 7 | + 10 |
| 49 | 50 |

Hey, *The Fudge Factor* triumphs again! By rounding, we were only one away from our actual total. Again, it will not always come out this close to the actual total, but rounding works great in a pinch.

Now, we will take a look at another way of making numbers behave.

## *The Pick 'n Shuffle*

Another neat way to add things up in a hurry is to pick the numbers apart and reassemble them.

Probably the most basic thing about numbers is that they are just convenient groupings for things, whether fish, dollars, or pieces of coal.

Let us say that the number 236 represents the number of paperclips in a pile on your desk. Now, there is nothing sacred about that number, 236. People generally give a total when talking about a number of things that they have, but it is just for convenience. It is all in how you decide to group them.

We could say that you have 236 paperclips, or we could say that you have 100 paperclips + 100 paperclips + 30 paperclips + 6 paperclips, or that you have 172 paperclips + 64 paperclips. The total number of paperclips does not change, only the way we look at them.

One very important thing to keep in mind when adding is to keep your columns lined up. This is generally done by lining up the "ones" columns of all the numbers you're going to be adding, but there's no shame in tacking on zeroes to the front of a string of numbers if you feel more comfortable aligning them on the left. All those extra zeroes say, after all, is "We do not have any thousands, nor any hundreds, but we do have some lovely tens and ones to show you."

A puzzle where you need to add numbers of different magnitudes, like 9602, 23, 707, and 14 might well look like this:

9602
0023
0707
0014

You are not getting graded on this; add the extra zeroes if you want. Even if you were taking a test, I cannot imagine that any teacher would take points off for setting up the problem in this way.

When adding decimals, like money, you line up all the decimal points, rather than using the ones column. Remember that $123 is exactly the same as $123.00; it is just that with the second number, we are reminding the world that we don't have any tenths or hundredths, while in the first example it is understood.

In school we were taught to add from right to left, "carrying" any numbers we had left over into the next column. The system works, but it is hardly the most efficient way—as you are going through, adding right to left, you really don't have any idea how big your answer is going to be. If you are doing the addition in your head, you are forced to remember the number backwards, as you get the last digit first. Not very useful, to say the least.

Adding from left to right tells you immediately at least how big the answer is going to be, and no lower. The further you carry the calculation, the closer you're going to come to the correct answer, depending on how much time you have and how accurate you need to be. And, of course, you find out the value of the first digits first, and last digits last, so it is far easier to keep track of than if you are doing the calculations mentally.

Just as we learned in Making Tens, it is generally more convenient to add round numbers (like 10, 300, or 6,000,000) than non-round ones. It is also easier to add small numbers than bigger ones. Suppose you had the following column of numbers to add:

82344
26221
99434
62382
14892

It may look daunting, but there is no reason we cannot bust these numbers up a little bit in order to make them behave themselves. Why not pick them apart and make them look like this?

| | | |
|---|---|---|
| 82344 | Means: | 80,000 + 2,000 + 300 + 40 + 4 |
| 26221 | Means: | 20,000 + 6,000 + 200 + 20 + 1 |
| 99434 | Means: | 90,000 + 9,000 + 400 + 30 + 4 |
| 62382 | Means: | 60,000 + 2,000 + 300 + 80 + 2 |
| 14892 | Means: | 10,000 + 4,000 + 800 + 90 + 2 |

Numbers like 90,000 may seem impressive, but they mean nothing more than that you have 9 of something, in this case units of 10,000 apiece. Now what? Well, depending on the accuracy you need, we can do a few different things.

The first thing we know, by adding up the number of units of 10,000 (notice we can find two "tens": 1 + 9 and 2 + 8, leaving us with a 6 to tack on to the end), is that there is no way we're talking about less than 26 units of 10,000 pieces, or 260,000. This may be close enough to a firm answer for you, but if not, we will go next door to the numbers representing 1,000 pieces.

We seem to have 23 sets of 1,000, which makes 23,000 or 20,000 + 3,000. Add 2 to the 10,000 place and leave the 3 where it is, which tells us that our answer is now going to come up to at least 283,000. Still not close enough? Okay, we'll see how many sets of 100 we have.

3 + 2 + 4 + 3 + 8 is 20, so we have 20 units of 100 or 2,000. Add the 2 to the 1,000s column and leave a 0 in the 100 column, as we have no remainder after the thousands are taken care of. Now we know our answer is at least 285,000.

Carry on with the 10s and 1s columns if you want to, but bear in mind that everybody fudges. In Physics class, our standard was to get within 10 percent of the right answer (which in this case would range from 256,500 to 313,500), but even if you were looking for 1 percent accuracy, the range of our answer would be from 282,150 to 287,850 and we already know the answer is not lower than 285,000.

There is no way that a bunch of puny 10s and 1s are going to add up to 2,850 in this case, as there are only five numbers in each column. Even if the remaining numbers were as high as they could be, which is 9 times the value of the column, we'd only have (5 • 90) + (5• 9) = 495, so the farthest we could be from a correct answer is 495, or a total possible margin of error of less than .2 percent. In many real life situations, you will be lucky to get a hold of numbers with a margin of error of 4 percent, or 20 times our highest possible error here, so we are really doing pretty well.

Notice that I used parenthesis in an equation in that last paragraph, so I suppose I had better explain what they mean before we go any farther. When I said (5 • 90) + (5 • 9) = 495 earlier, what I meant was, "Take 5, multiply it by 90, then take another 5 and multiply it by 9, and then add the whole thing together." More on parentheses when we get into more advanced stuff.

I used "•" as a sign to multiply there, which is the sign people who work with computers use to say "multiply." Most people are used to "X" as a sign for multiplication, but "X" the sign can get confused with "x" the variable when we get into algebra. Generally, I'll be using either "•" or parentheses to indicate multiplication, (4)(5) = 20.

Of course, with the *Pick 'n Shuffle* method, I do not expect you to take the time to write down the 10,000s, 1,000s, 100s, and so on separately. You can if you want to, of course, but it is a lot faster if you just add single digits, left to right, keeping in mind things like, "Okay, I am adding 10,000s now, so that what I have here turns out to be a total of 80,000."

Another way to shorten time spent on addition or subtraction of two numbers is to add or subtract enough from one of the numbers to make it a nice round one, then do the reverse to the answer (add if you subtracted, subtract if you added).

What's the point? It is a lot easier to add 123 to 100 than to add 117 to 94. Sounds confusing, I know, but it could not be simpler. Here is an example of what I am talking about:

```
 117     117          117
+ 94    + 94 + 6    + 100
                      217
```

Once you change one of the two numbers to a nice round 100, the problem's answer is obvious. But since we added 6 to the bottom number, we need to subtract 6 from the final answer of 217, leaving us with 211.

Always pick the more convenient of the two numbers (that is, the number closer to a nice round number like 10, 50 or 100) to modify; that way it does not get overly complicated when you add or subtract the modifier number from your answer. Again, this method works well for addition and subtraction, but not multiplication or division. This is why:

```
 117     117          117
• 94    • 94 + 6    •100
```

Sure, it is easier to multiply 117 by 100; all we have to do is add two zeroes to 117 for 11,700. But then we need to get rid of our modifier—instead of just adding or subtracting 6 to the problem, though, we ended up multiplying 117 an extra 6 times! We have to figure out what 6 • 117 is and subtract that. Kind of a mess, really.

## *Checking your work: Addition*

In grade school and high school, my teachers always made us check our work. Why they didn't just make it easier to do math in the first place is beyond me—maybe someone else chose the textbooks.

It was always kind of a hassle, this checking my work. As soon as they stopped demanding it, I stopped doing it. But then I learned about a really simple and almost foolproof way to check one's work. Truthfully, I still do not check my work very often because I have faith in my math skills, but I will present the technique for those who wish to use it.

In order to check your work on addition problems, add up all the digits in each row (rows go side-to-side, columns go up-and-down) and see what you get. If you get a two digit number, like 17, add the 1 and the 7 to make 8. If you end up with 9, treat it as a 0. Write the numbers down at the side of each row or just remember them. Then do the same for the answer. So you might have a puzzle that looks like this:

```
   2238      2238     6
   1490      1490     5
   8394      8394     6
 + 1120    + 1120     4
  13242     13242     3
```

Now add all the check numbers from the rows in the same way and compare them to the digit you wrote down next to the answer. If they're the same, the chances are really good that you've done the puzzle correctly. In the above example, 6, 5, 6 and 4 add up to 21, or 3, which is the same number as we got for the answer, so it is probably correct.

This system is not foolproof—it won't help if you've transposed two numbers in a row when writing them down, and there is always the off chance that you've goofed up so dramatically that you've come up with another number which has a check digit that equals the sum of those in the puzzle, but by and large it works pretty well. See future chapters for checking your work with subtraction, multiplication, and division.

Turn to the "addition" chapter in the workbook, and complete the puzzles until you are sure you know what you are doing. When you have made addition your own, come back and learn some shortcuts for multiplication.

# Multiplication shortcuts

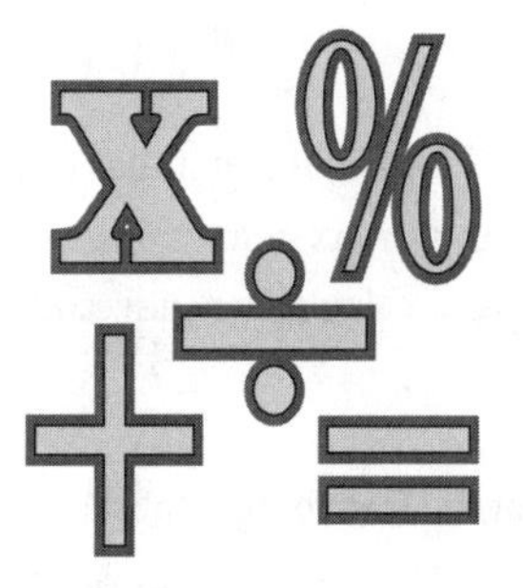

# Chapter 3
## Multiplication shortcuts

## Shortcuts that work

We are jumping from addition to multiplication because most people find it easier to add stuff up than take it away, and multiplication is really just a shorthand method for adding anyway.

When I say "four times five equals twenty," what I really mean is "take four sets of five somethings, add them up, and you get twenty." In mathematical terms, $4 \cdot 5 = 5 + 5 + 5 + 5$

You probably have some of the multiplication tables memorized from grade school, at least for the numbers up to 10. While I do not lean heavily toward rote memorization, I believe that having the multiplication tables for the numbers 1 through 10 memorized is an absolute necessity for quick multiplication skills. If you are a little rusty, you can pick up some flashcards in just about any bookstore, or make your own. Remember, the time you spend practicing is not wasted; rather, it is an investment in yourself and your free time in the future.

One of the tricks for multiplying numbers in your head is to bust them up, just like we did with addition.

Friends of mine who are messing with paperwork or otherwise deep in thought often call across the room to me, "What's 21 times 18?"

"378," I yell back after thinking about it for a little bit. It is not that I am known across North America for having super math powers or anything; it is just that my friends are lazy and they do not have a calculator handy, and I enjoy doing math on that chalkboard in my head.

When somebody asks me a question like that,the first thing I do is factor the two numbers, because it is much easier to multiply a number by a one-digit number than a two- or three- digit number. By "factoring," I mean that I think about how to break at least one of the numbers into its simplest parts. Can I divide the number by 2? By 3? By 4? Here's how to tell:

- A number is divisible by 2 if it is even; that is, if it ends in 2, 4, 6, 8, or 0.

- A number is divisible by 3 if the number's digits, added together, equal 3 or a multiple of 3. Example: Is 27 divisible by 3? Yes, because 2 + 7 = 9, a multiple of 3. Is 127? No, because 1 + 2 + 7 = 10, which doesn't divide evenly by 3.

- A number is divisible by 4 if it is even and its last two digits are evenly divisible by 4. 1,632 is divisible by 4 because 32 is.

- A number is divisible by 5 if its last digit is a 5 or a 0.

- A number is divisible by 6 if it is divisible by both 2 and 3.

There are rules for dividing numbers by 11, 13, 20 and so on, but the vast majority of numbers will be covered by the five rules above.

Don't bother dividing numbers by 1, by the way, as every number is divisible by 1 and it just puts us back where we started. That is, 237 / 1 = 237.

In a similar vein, dividing by 0 is not allowed.

Once you can divide the number by 2, 3, 4 or 5, try it again to see if you can break it up further.

Back to our original puzzle, 21 • 18:

21 multiplied by 18 is the same as 3 • 7 multiplied by 2 • 3 • 3, but since I do not want to multiply both 3 and 7 by four different numbers, I would rather think about it as 3 • 7 times 2 • 9.

Now then: It just so happens that 3 • 7 • 2 • 9 gives us the same number as 21 • (2 • 9) or (3 • 7) • 18 would, but I do not know my multiplication tables up to 18, so I am using smaller numbers and performing more steps.

I generally multiply the larger numbers first, so that when the numbers get big, halfway through, I only have to multiply them by 2 or 3. I will multiply 7 and 9 first, getting the pesky big numbers out of the way. I know they make 63. Looking at my factors, I see that I only have to multiply 63 by 2 and 3 to be done. I will take the larger of the two first.

Luckily, (3 • 60) + (3 • 3) is the same as 3 • 63, so I can break it up further by using subtraction to round out that clumsy number.

3 times 60 is 180, pretty easy, then add 3 times 3 for 9 more, so 3 times 63 is 189. I still need to double it to get the final answer though (2 times 180 is 360, and 2 times 9 is 18, so the total is 378.)

It may sound a little confusing, but we'll work on that. Just remember these two rules:

1) If splitting numbers up as factors, take the two biggest factors and multiply them together first. Then multiply that result by every remaining factor, in descending order of size.

2) If splitting up numbers by subtraction (i.e. 63 – 3 = 60) remember that you have to multiply both parts by any remaining factors.

Let's try another one: 32 times 45. Both have plenty of factors, so it should be a snap. I will take 2 • 2 • 8 times 3 • 3 • 5 so as to keep our factors as low and few as possible. First we multiply the biggest factors, 8 and 5, to get 40. Next we look around for our next highest factor, which happens to be a 3. 3 • 40 is 120, a nice round number.

Our next highest factor is also a 3, so we will run that through the system and get 360. Now we are only left with two 2s, so the rest should be easy. 2 • 360 is 720, which is not too tough, and our last 2 times 720 is 1440, even easier. Here is how it looks on paper:

```
  8     40    120   360   720
 x 5    x 3   x 3   x 2    x 2
 40    120   360   720  1440
```

## Multiplying by 5, 10, 15, etc.

Multiplying any number by 10 is easy—just add a zero to the original number. When multiplying by 10, 30 becomes 300, 423 becomes 4230, and so on.

To multiply by 20, start by multiplying by 2, then add a 0 to the final answer. 24 becomes 48(0), or 480, 56 becomes 1120, and 36 becomes 720.

Multiplying by 30 is the same deal—multiply instead by 3, then add a 0.

You can multiply numbers by 100, 200, and 300 in the same way. Just add two 0s to the end of the answer, instead of one. How much is 1,000,000 times 476? Count the zeroes in a million; there are six of them, so multiply 476 by 1 (leaving 476, obviously) and add six 0s.

Multiplying by 5, 15, and 25 is not much tougher. It all comes from playing with numbers. Multiplying a number by 5 is just like doing it by ten, except after adding the final 0, you divide the answer by 2. Thus, 24 • 5 starts out as 24 • 10 = 240, but then you split it in half to make 120. Again, 362 • 5,000 (think 10,000 so you add the right number of zeroes) becomes 3,620,000 then 1,810,000 when you split it in half. Remember, just because 3,620,000 has more zeroes than 362, don't let it intimidate you. All those zeroes do is fill up space in the thousands, hundreds, tens, and ones columns, and let us know those columns are empty.

To multiply by 15, multiply first by 10 and then add half of it to the original answer. This gets a little trickier, but it is nothing that cannot be done in your head if you keep your wits about you. 45 • 15 starts out as 45 • 10, or 450, and then you add half of it, or 225, for a final answer of 675. Are you starting to guess how we will multiply things by 25?

Right! Multiply by 10, remember what half that number is, double the one you multiplied, and add the half that you were remembering. It is actually simpler than it sounds. Start out with a number, say 234, multiply by 10 to get 2340, and figure out half of that, which is 1170. Tuck that number away for future reference. Now double 2340 to make 4680, and add the 1170 you stored earlier to make 5850. You do need to be a little careful to do your addition right, but it is a lot faster than working it out longhand.

Since there are four 25s in 100, you can also always multiply numbers by 25 by first multiplying them by 100 (or tacking on two 0s), then dividing by 4. Thus, 234 becomes first 23400, then 5850 when we divide by 4.

Multiplying numbers by 50 is even easier. Just multiply by 100, then cut the answer in half. 440 • 50 becomes first 44,000 then 22,000 when you cut it in half.

Remember that 200 • 423 is exactly the same as saying 423 • 200, so it does not matter which part of the equation our nice round number is. Ready for more?

Whenever you multiply a number by 11, 22, 33, or any number that has both digits the same, you can simply multiply the number by 10, 20, 30, or whatever, then add 10 percent.

Dividing by 10 works much the same as multiplying by 10, except you take away a 0 instead of adding one. Consider 227 • 33. Multiply 227 by 3 first, which is a little tricky, but not too bad, it is really only (3 • 200) + (3 • 20) + (3 • 7), which gives us 681, tack a 0 on the end for 6810, and add 10 percent, which is...hey! It is 681, the same number we got when we multiplied by 3! That gives us 7491.

When working with bigger numbers, remember to keep your columns in line when adding. If it helps to think of 681 as 0681, just so the columns match up, feel perfectly free to do so.

When multiplying decimals, treat them just like regular numbers until you get your answer. Then, count over, right to left, on your answer and put down a decimal point where you have counted the same number of places as you have decimal places in the whole problem.

It is easier than it sounds: 1234.56 • 43 comes out to be 5,308,608 when we multiply it (no trick for that one, I had to do it longhand), but we have two numbers to the right of the decimal point in the original equation, 5 and 6, so we count two numbers to the left from the ones column and put down a decimal point, giving us 53086.08 as a grand total.

## *Exponents and all that*

Once in awhile, usually in a scientific application, you'll come across numbers like $3^5$ or $10^6$. These are nothing but shorthand ways of saying, "Take a 3, and multiply it by itself 5 times." The tiny number is called an exponent, a word I will continue to use because it is more convenient than saying "the tiny number to the upper right of the big one."

Thus, $3^5 = 3 \cdot 3 \cdot 3 \cdot 3 \cdot 3$, or 243. In the case of $10^6$, you could multiply 10 by itself 6 times, or you could just write down a 1 and add six 0s.

Because so many numbers in science are so ungodly large or small, it often makes sense to express them as $1.234 \cdot 10^6$, rather than 1,234,000. If you come up against a decimal like that, rather than add six 0s, you just move the decimal point to the right six places, adding 0s when you run out of numbers.

If you should happen to come across a negative exponent, like $14^{-23}$, all the negative sign means is that you need to move the decimal point to the left 23 places, rather than to the right.

A couple of things about exponents which may seem strange at first:

1) Any number, n, with an exponent of 1 has a value of n, because the exponent isn't high enough to provide it with copies of itself to multiply itself with.

2) Any number, n, with an exponent of 0 has a value of 1, except 0, which is always 0 no matter what you multiply it by. The justification for this seems to be (according to mathematicians) that if $5^1 = 5$, and $5^{-1} = .5$, then $5^0$ should be something in between, but why they picked 1 is beyond me.

## *Checking your work: Multiplication*

Checking your work with multiplication is much the same as with addition, except you multiply the check digits from the two numbers instead of adding them. Then compare them against the check digit of your answer.

Example:

| | | |
|---|---|---|
| 424 | 424 | 1 |
| x 23 | x 23 | 5 |
| 9752 | 9752 | 5 |

1 • 5 = 5, so it looks like we multiplied correctly.

Now we will leave multiplication alone for a bit; work the puzzles in the second section of the workbook until you feel sure of yourself, then come join me for tips on subtraction.

Subtraction shortcuts
4

# Chapter 4

## Subtraction shortcuts

### *Or, addition in reverse*

Many people find subtraction marginally more difficult than addition; for some reason, it just feels more right to add than subtract. But what if you could apply the ease of addition to subtraction problems?

Scott Flansburg, another "math made easy" author, calls subtraction "addition in reverse," and he is completely correct. When we get into playing with negative numbers, at the end of this chapter, it will become clear to you that adding negative numbers is just like subtracting.

Do you remember how your teacher told you to check your subtraction problems, to see if they were correct? Say you just finished working a problem like this:

$$\begin{array}{r} 227 \\ -\,103 \\ \hline 124 \end{array}$$

In order to check your subtraction, you were probably told to add the answer you got to the number which was being subtracted; that is, to add 124 to 103. This simple little technique is the key to easy subtraction.

To subtract any number from any other number, just figure out what number, added to the one being subtracted, makes the original number. Using mathematical symbols, then, subtraction is less a question of 234 - 102 = ? than of 102 + ? = 234. As Ross Perot would say, "It's just that simple."

Since subtraction is nothing more than rearranged addition, it makes sense that we can steal a few strategies from the addition section. Subtraction usually only involves two numbers (not counting the answer), so Making Tens is not all that useful, but we can certainly round the numbers, and we can also speed things up by working left-to-right.

We can also add and subtract from the numbers being subtracted in order to make them easier to work with. In the above example, 234 – 102, we might subtract 2 from 102 in order to simplify things as long as we remember to subtract 2 from the other number. Otherwise, the answer would be 2 too high.

For larger numbers, a left-to-right, pencil and paper version of subtraction works pretty well. The chances are much better that, in a bar or on an airplane, you will have an easier time borrowing a pen than a calculator.

Subtract from left-to-right, just like we add. When you get to a place where the digit which is being subtracted is larger than the one it is being taken from, pretend the number on top is actually 10 more than it is, and subtract normally. Write the result down, then make a slash across the number before it. It works like this:

| | |
|---:|---:|
| 4288621 | 4288621 |
| - 1194203 | - 1194203 |
| | 3194428 |

When you look at your answer, "see" each number with a slash through it as one less. Hence, the answer above would actually be 3094418, rather than 3194428. After practicing for a while, it is likely that you will "automatically" see numbers with a slash through them as one less. Pretty slick, eh?

## ***Checking your work: Subtraction***

To check your work with subtraction, you add the check digit of the answer to the check digit of the number being subtracted to see if it adds up to the check digit of the number being subtracted from.

Example:

```
  100     100    1
 - 47    - 47    2
   53      53    8
```

Since 8 and 2 make 1, this puzzle checks out. Remember, 8 + 2 = 10, then add the 1 and 0 of the ten together, 1 + 0 = 1, which leaves you with that great single digit check number.

## Negative numbers

In our universe, there exists a type of matter called "antimatter," which is exactly like regular matter, except the electrical charge of its atoms is reversed. An antimatter sofa, for example, would look just like a regular sofa, weigh the same as a regular sofa, and, to a person made of antimatter, would be just as comfortable. Should some antimatter ever come into contact with regular matter, however, both types of matter would immediately release all of their energy in a tremendous explosion.

Negative numbers are much the same as regular numbers. They have the same value, but instead of being the proud owner of 200 peanuts, a negative number indicates that you owe 200 peanuts.

Negative numbers are what you get if you subtract a larger number from a smaller one. If a negative number and a positive number of the same value are added together, they'll add up to 0. Think of it this way:

On the night before you get paid, you need to fill the car with gas, but doing so reduces your checking account balance to – $18. When you deposit your $1,000 paycheck the next day, you add it to your check register, but since you're $18 below zero you have to subtract $18 from the $1,000 deposit. That's all there is to adding negative numbers.

Subtracting negative numbers is a little trickier. When you subtract a negative, you're really adding a positive. Consider this example:

If you owed me $10 and wrote me a check for it, you would subtract it from your account balance. I am such a nice guy, however, that when I receive the check, I decide to tear it up. When I tell you what I did, you feel comfortable adding that $10 back into your account. What you are really doing, however, is subtracting a negative number.

Multiplying and dividing negative numbers is not too tough, either. Just remember that in multiplication and division, negative numbers poison everything they touch. That is, whenever you multiply or divide by a negative number, your final answer will be affected. When you need to work a puzzle like: 45 • – 4, your final answer will be a negative number, –180. Why? Because what you are actually saying is, "I need to add together these numbers: (– 4) + (– 4) + (– 4) ..." and so on. Obviously, then, your answer will be negative.

If multiplying a negative number by a negative number, though, the answer will be positive. (– 45) • (– 4) would be 180, because you're taking away 45 sets of 4 somethings owed.

Division involving negative numbers is exactly the same. If you have 1 (or 3 or 5) negative numbers in your equation, the answer's going to be negative. If there are an even number of negative numbers, the answer will be positive, because 2 negative numbers who come into contact will make a positive one, just like in multiplication.

Other than that, all operations done with negative numbers are exactly the same. As long as you're careful to keep the signs straight, you'll be all right.

Turn to the workbook now for some practice with subtraction and negative numbers, then come back to shortcuts the boogeyman of basic arithmetic, long division.

# Long division shortcuts

# 5

# Chapter 5

## Long division shortcuts

If you'll remember, long division in school was always a mess. Huge problems sprawled across the page, with tails that dangled six inches or more. Each problem took 5 or 10 minutes to work. We're about to change all that.

Division is really a two-step process. The first step is guessing how many times one number fits into part of another. The second part is subtracting one number from another. If we are faced with the following equation, for example:

$$22\overline{)1234}$$

The first thing we need to do is to decide how many times 22 goes into 12. None, obviously. If that doesn't work, we need to add another digit and see how many times it goes into 123. Using rounding in conjunction with a little mental multiplication, you can see that the first digit of the answer is going to lie somewhere between 5 and 7. On the low end, 5 • 20 is 100, while 7 • 20 is 140, too high.

If we pick 22 apart, into 20 and 2, we can see that (5 • 20) + (5 • 2) = 110. Is there room for another 22 in 123? No way—that would put us up to 132. So the first number in our answer is 5. We subtract the total of 5 • 22 from 123, left to right, and get 13. 22 obviously doesn't fit into 13, so we bring down the next digit, a 4. 22 does fit into 134, which coincidentally enough is pretty close to what we got when we multiplied 6 and 22, or 132.

We subtract 132 from 134, leaving us with 2, and we see that there aren't any more digits to bring down, so we're essentially done. Our answer is 56 with a remainder of 2, or 56-2/22.

Using the *Everyday Math* principles of picking numbers apart and subtracting left to right, there's really no reason to write anything down on paper. Just keep tacking on numbers to the end of the answer and remember which number you need to bring down next. Of course, you don't even need to work the problem fully. If your Fudge Factor allows, you can say that the answer is about 56, dropping the remainder altogether.

Feel free to round the numbers in the equation, as well. We could have asked what 1230 divided by 20 was, which would have been simple enough to do without setting it up formally. We could have just marched across, left to right, and gotten 6, 1 and 1/2, or a final answer of 61-11/22.

Remember that as far as division or multiplication is concerned, you can multiply or divide both numbers involved by any number as long as you do the same thing to the other number. The answer will be the same as you would have gotten anyway. Remember that multiples of 10 are particularly easy to work with. Just knock a 0 off each number. 120 divided by 20 is 6, just like 12 divided by 2.

In the above example, we couldn't divide both sides by 2, 3, 4 or 5, but you'll find that many times you will be able to do so. Look at the puzzle below:

$$420\overline{)142260}$$

What larks! We can divide both numbers concerned by 2, 3, 4 5, and 10! Where shall we start? Generally it is easiest to divide by higher numbers first, so we will divide each number by 10 and see where that leaves us:

$$42\overline{)14226}$$

Better, but it can be reduced even farther. We cannot divide by 5 anymore, because neither number ends in 5 or 0, nor can we divide by 4, because 42 is not evenly divisible by 4. We can divide by 3, though, which leaves us with 14 doing the dividing and 4742 being divided.

$14\overline{)4742}$

Now what? Well, we can still divide each number by 2, giving us:

$7\overline{)2371}$

Which is about as reasonable as we can expect this problem to become. We will have to work the rest of the problem left to right, and just by looking at it we can tell that the first digit of the answer will be 3.

That leaves a remainder of 2, so we drop the 7 down to meet it. 7 goes into 27 a total of 3 times, so the second digit of the answer is also a 3. When we subtract 21 from 27, we are left with 6, or 61 when we drop the 1 down.

It is easier, in this case, to determine what 7 • 10 is and subtract from that, rather than keep multiplying larger and larger numbers by 7. 7 • 10 is 70, but that is too high. What about 7 • 9, otherwise known as 70 - 7? 63, still too high, but not by much.

Obviously, then, the last digit in our answer will be 8, with a remainder of 5, for a grand total of 338, remainder 5.

## Division beyond remainders

But "remainder 5" is such an ugly way to end our answer. We could leave the answer as 338 and 5/7, but that too seems unfinished. Why don't we convert that remainder into a decimal? When we have exhausted the possibilities of dividing the original number by 7, all we need to do is put a "." after 338 and 2371, then pretend that there is an infinite string of 0s behind 2371. There is, after all, because 2371 is exactly the same number as 2371.000000; it is just written down differently.

Usually, we can just keep dividing until we finally end up with an answer which divides evenly. Unfortunately, we have no idea how long this calculation will take; we may have to keep dividing through 50 zeroes in order to come to our final answer, or it may never happen.

It is for reasons of convenience that I recommend that we stop after the "hundredths" place in the answer. Most answers really do not need to be any more accurate than that. Let's see what happens:

```
    338.
7)2371.00
      50
```

7 • 7 is 49, which is 1 less than 50. Hence, our answer is now 338.7. When we bring the next 0 down, we have 10, so our last digit in the answer is 1. At this point we can either look at the next digit, to see if we should round 1 up to 2, or we can just stop, and "truncate" the answer. Let's take a look at what we would have if we paid attention to the "thousandths" column.

If we had continued to bring down 0s, our next number would be 3, which is not high enough to force us to change the 1 into a 2. Our final answer is thus 338.71.

## Fractions

Fractions are really nothing more than unsolved division problems. When you see a fraction like $^{11}/_{22}$, all it really means is that it represents the answer to 11 divided by 22, but that it was more convenient or time-effective to leave it in its present form.

Perhaps it's just me, but whenever I see a fraction, I think somebody somewhere is goofing off—decimals are much easier to handle, and if they had just divided the darn thing through. To get a "real" number out of a fraction (generally a decimal, except in math books), simply divide the top number by the bottom.

What if you come up against something like $5\frac{3}{4}$ (a "mixed" fraction)? Well, whenever you have the same number on top of a fraction as you do below, such as $\frac{4}{4}$, it means 1. To make a decent number out of a mixed fraction, all you do is multiply the "whole" number (5 in this case) by the bottom number in the fraction and add it to the top part of the fraction. Then, divide the whole works by the bottom number. Thus, $5\frac{3}{4}$ becomes first $\frac{20}{4}$ + $\frac{3}{4}$, or $\frac{23}{4}$, then we divide through to get 5.75.

Look at the previous paragraph. Did you notice the same thing that I did? We could have just left the whole number the way it was and simply divided 3 by 4 to get the decimal part. In that case, we would have been fine, but it is important that you know how to do it, because some puzzles will call for converting 5 into fourths, elevenths, or sixty-thirds.

I also snuck one over on you; I added $\frac{20}{4}$ and $\frac{3}{4}$ without explaining fraction addition. It is pretty easy, however. To add fractions, just make sure the number on the bottom is the same. If it is not, like $\frac{5}{3}$ and $\frac{2}{6}$, you will need to do some quick multiplication to make it so.

Remember that $\frac{2}{3}$ and $\frac{4}{6}$ come out to be the same number, or percentage of a whole piece of something. $\frac{1}{2}$ of a pie, after all, is exactly the same as $\frac{2}{4}$ of a pie. With this in mind, you know that $\frac{5}{3}$ and $\frac{10}{6}$ come out to be the same number, so you could just double $\frac{5}{3}$, top and bottom, in order to be able to add it to $\frac{2}{6}$. It turns out that $\frac{2}{6}$ is the same as $\frac{1}{3}$, so you could have changed either part, in this case.

Sometimes when you are working on a puzzle which has the same variable both on top of and on the bottom of the bar signifying "divide," (it is really the same bar as in fractions, just longer), you can cancel the two variables out, since you are both multiplying by and dividing by whatever that variable is. This is really just an extension of simplifying fractions. Here is an example:

$$\frac{4x}{12x}$$

We could cancel out the xs on both top and bottom, since it is really like saying $\frac{4}{4}$, or 1.

What if you are stuck needing to add two fractions like $\frac{3}{7}$ and $\frac{2}{5}$? I would just turn them into decimals and add the decimals, but every math book in the world seems to think it is important for you to know how to add them in fraction form.

To add unruly fractions like $\frac{3}{7}$ and $\frac{2}{5}$, you need to seek the lowest common denominator . The denominator is actually what the number on the bottom is called, and the "lowest common" part of the phrase means that you want to find the smallest number that can be divided evenly by both 5 and 7.

Sometimes (as in this case) you will just end up having to multiply the numbers to get the lowest common denominator (35). When you turn $\frac{3}{7}$ into something over 35, you multiply 7 by 5, so do the same on the top half to get $\frac{15}{35}$. To make $\frac{2}{5}$ into something over 35, you need to multiply 5 by 7, so multiply 2 by 7 as well. Add the two fractions, and we get the grand sum of $\frac{29}{35}$, which is as ornery a fraction as I've ever seen. Sorry. Using the *Fudge Factor*, however, we can see that $\frac{29}{35}$ is pretty close to $\frac{30}{35}$, which reduces (when we divide both the top and bottom numbers by 5) to $\frac{6}{7}$.

To subtract fractions, you must also have a couple of fractions with the same denominator. Then just subtract the second top number from the first. $\frac{4}{5} - \frac{1}{5} = \frac{3}{5}$.

Multiplying and dividing fractions is a different story. To multiply fractions, multiply the top number by the top number, and the bottom number by the bottom number. No common denominator is needed. $\frac{2}{5} \cdot \frac{4}{7} = \frac{8}{35}$.

To divide fractions, you need to cross-multiply, or multiply the top number of one by the bottom number of the other, and vice versa. $\frac{2}{5} \div \frac{4}{3} = \frac{6}{20}$. It's a strange system, but it works.

## Decimals and percentages

The word *percent* means per hundred. That is, if a commercial says that 66 percent of doctors prefer Grinno toothpaste over the leading brand, 66 out of 100 doctors surveyed apparently feel this way (or so the advertisers would have us believe—more on statistics in a few chapters). This number can be displayed in ratio form, as 66 : 100 or as a fraction, $^{66}/_{100}$.

As we learned when "finishing" fractions to make decimals, $^{66}/_{100}$ divides out to be .66. All that is required to make a percentage out of a decimal is to move the decimal point two spaces to the right and add the "%" sign.

You can add and subtract decimals without doing anything to them. That is, if 24 percent of customers think your company is the greatest thing on earth and 35 percent think it's merely pretty good, you can say that 59 percent of customers surveyed like the company.

To multiply and divide percentages, however, you must turn them back into decimals by moving the decimal point two places to the left. Example: Lazy Larry just found out he's getting a 6 percent raise. If his current salary is $60,000 per year, what will his new salary be?

$60,000 (current salary) • .06 (raise) =
3600.00 more—or a total of 63,600

## Square roots

Every once in awhile, you will need to find the square or cube root of a number. Remember that in chapter 2, we talked about how we could use a shorthand method of saying "multiply this number by itself." $5^2$ means the square of 5, and $5^3$ means the cube of 5, and so on.

When we say we are looking for the square root of 25, what we are saying is, "Find me a number which, when multiplied by itself, will equal 25." Obviously, the answer in this case is 5.

Because nice, round squares are somewhat rare (there are 10 between 1 and 100, and a lot less per hundred after that), working out square roots of numbers will tend to be a messy business, full of decimals. In most cases, it is generally faster to use a calculator, but in case you do not have one, there are a few tricks available.

Try comparing the number to the nearest squares. If you need to figure out the square root of 70, for instance, remember that the square of 8 is 64 and the square of 9 is 81, so the square root of 70 lies somewhere in between. In this case, 70 is 6 away from 64 and 7 away from 81, so the true square root would be a little under halfway, maybe 8.4 or so. I will check it with a calculator: $(8.4)^2$ is 70.56; about as close as we will come without resorting to large strings of decimals.

If you have a pencil and paper handy, there is a somewhat more involved way of finding square roots to a greater degree of accuracy. It still involves the Fudge Factor, but if close is good enough, you may wish to use it.

Suppose you need to find the square root of a number like 85. You know that 9 • 9 is 81, too low, and 10 • 10 is 100, too high. Obviously the answer is going to be "9 something", so write down a 9. To get the rest of the answer, divide 85 by 9, which gives us 9.444444444 and so on, which we will treat as 9.44, just dropping the rest of the numbers because we round down (or truncate) for 4, and round up for 5 or more.

Next we add 9 and 9.44 and average them to get 9.22, which is our final guess at the answer. In fact, $(9.22)^2$ is 85.0084, so we are really pretty close.

These techniques work well for stage magicians, who often have people yelling perfect cubes up at them, but there are only 4 perfect cubes between 0 and 100 (1, 8, 27, 64), and they get a lot rarer after that. Should you ever need to find the cube root of a number, the chances are very good that it will soon degenerate into messy strings of decimals. You are better off punching it into a calculator.

## *Checking your work: Division*

To check your work for division, multiply the check digit of the answer by the check digit of the number doing the dividing. If you come up with the check digit of the number being divided, you're all set.

# More valuable shortcuts

# 6

# Chapter 6

## More valuable shortcuts

Math is cool—there is a standard order in which to do all of the adding, multiplying, etc. This standard is called "the order of operations," and was designed so that two people working the same equation could come up with the same answer, at least as long as they did all the number–crunching right. It does make a difference in most cases. If I gave you the monster equation below and said, "Okay, $x = 3$, go to it," you might a) go into shock, or b) begin anywhere and work your way around, thus getting the wrong answer.

$$\frac{(47x^2 + 32x)^2 - (12x^3 - 4x^3 + 12)^3}{43 - x}$$

But thanks to the order of operations, you'll soon know where to begin. The order of operations is as follows, and it generally proceeds from left to right, top to bottom:

1) If you have the value of $x$, substitute it for $x$ everywhere in the problem.

2) Work inside parentheses first.

3) Resolve exponents and square roots—for example change$3^3$ to 27.

4) Work multiplication and division within parentheses.

5) Work addition and subtraction within parentheses.

6) If parentheses have exponents on them, resolve those.

7) Multiply and divide parentheses above the "division" bar.

8) Add and subtract parentheses above the bar.

9) Repeat steps 1 through 8 below the bar, if required.

10) Divide the top by the bottom, if required.

I know it sounds like a lot of work, but most puzzles you're likely to come across will not have nearly this much to do. I just want to make it clear to you that when you see a problem like: (5 + 3) • 4 + 2, first you need to add 5 and 3, then multiply them by 4, then add 2.

There is a neat little mnemonic enhancer for remembering the order of operations, generally: "Please Excuse My Dear Aunt Sally." If you can remember this sentence, you can remember the order of operations. The initial letters stand for:

Parentheses
Exponents
Multiplication
Division
Addition
Subtraction

Just remember a couple of things: First, plug in any variables you may happen to know the values of, and second, work generally from left to right, top to bottom.

Let's work a couple of problems for practice:

12 + (45 • 12) – (19 + 13) = ?

The first thing we do, since we have no exponents or variables to deal with, is to work inside the parentheses. Once we do the work inside the parentheses, we end up with:

$$12 + 540 - 32 = ?$$

Then we work left to right, adding and subtacting as necessary, and end up with an answer of 520.

Let's try another one with a little more complexity:

$$(x^3 + x^2 + 7) - (24 \div x) = ? \; x = 3$$

Now that we have some exponents and variables to play with, we had better work with them first. Since we know that $x = 3$, we first plug a 3 wherever we see an $x$:

$$(3^3 + 3^2 + 7) - (24 \div 3)$$

$3 \cdot 3 \cdot 3$ is the same as $9 \cdot 3$, or 27. $3 \cdot 3$, obviously, is 9. We will now substitute these numbers for the ones with exponents and proceed to work inside the parentheses.

$(27 + 9 + 7) - (24 \div 3)$ becomes $(43) - (8)$

Giving us a final answer of 35. There's no trick to it; just remember to "Please Excuse My Dear Aunt Sally."

But what if we had not been given the value for $x$? Would the puzzle have been unsolvable? Maybe, or maybe not. If you have an equation with an unknown quantity, you can still solve it as long as you remember the Golden Rule of Equations:

***"Doest Thou to One Side What Thou Doest to the Other"***

That is, whether you multiply, divide, add or subtract from one side of the equation, make sure you do it to the other side too. Let's start out with a simpler example:

$$3(x^2 + x) - 36 = ?$$

In this case, we really can't do anything with exponents or inside the parentheses, since we do not know what number $x$ is equal to. Therefore, we will have to skip those steps and come back to them when we have simplified the problem a bit.

We will start by making the whole equation equal to 0. Well, why not? We do not know a better number to make it equal to, and setting an equation equal to 0 has certain advantages. First of all, we can now divide both sides by 3—just a modification of what we did in the last chapter when working with division problems. Now we have:

$$(x^2 + x) - 12 = 0$$

Since 0 divided by 3 still leaves 0, we have simplified the problem somewhat. What else can we do? We can add 12 to both sides. When we get $x$ alone on one side of the puzzle, we generally get an answer.

$$(x^2 + x) = 12$$

Or, since we don't need the parentheses anymore,

$$x^2 + x = 12$$

Now what? We could simplify by turning $x^2$ into $x \cdot x$, leaving us with:

$$x \cdot x + x = 12$$

That's really as much as we can simplify it. Now we have to start plugging in numbers and thinking to ourselves "What number, when multiplied by itself and added to itself, equals 12?"

Since 12 is such a low number, the chances are good that $x$ is less than 5. $5^2$ is 25, after all, and adding another 5 makes it 30. Let's try 2. 2 • 2 + 2 = 6, so that is not it. How about 3? 3 • 3 + 3 = 12, so we have found our answer. Or have we?

What about – 4? – 4 • – 4 = 16, then when we add another – 4, we get 12. It turns out there are actually two answers to this equation! Until we find out more information about the puzzle, we are stuck with two answers.

If the puzzle described the number of autos which could be manufactured per day at a certain plant, the answer would obviously be 3, since a factory generally is not set up to make new automobiles disappear.

If the puzzle described the amount of debt a company paid off in an average day, though, the answer could be – 4. Generally, if you are making up equations to describe how something happens in your life, you will know which of the the answers you come up with is correct.

## Functions

Do you remember the introductory part of the book, when I showed you an equation like this:

$$f(x) = x^3 + 2x^2 + 4x; f(4)$$

It does not look so intimidating now, does it? That string of numbers and letters is really called a function. Functions are nothing but number factories. They have a manufacturing process all set up, and you can just run any number through them.

The first part of the statement, $f(x) = x^3 + 2x^2 + 4x$, does nothing but tell us what processes in the factory have been set up. That is, when you plug in a number, the factory will cube it, take two times its square, and multiply it by four, then add the whole mess together.

The second part, f(4), just says, "In this particular case, we want to plug in the number 4." That is all there is to functions.

What good are they? Why, they are very useful. If ever you need to convert kilograms into pounds , you can use a function: $f(x) = 2.2x$, where $x$ is the number of pounds you need to convert. If you ever need to barter zucchinis for tomatoes, you can use a function. Functions are really only slightly more permanent versions of equations.

There really is no shortcut to checking your work with equations like this, because we are mixing addition, subtraction, multiplication, and division. However, if you have it written down in front of you, or if you were good at visualization, it does not take much time to run through the problem again.

There are some puzzles in the workbook which await your attention. When you are sure that you know what you are doing, at least as far as the order of operations is concerned, come back to learn about solving story problems.

# More story problems

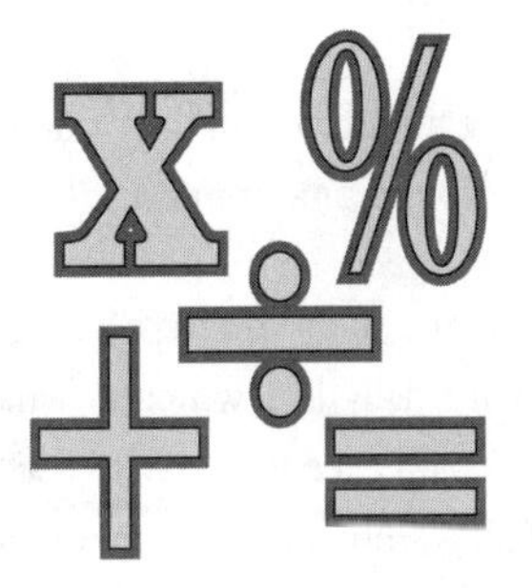

# Chapter 7

## More story problems

When I was a kid, I loved story problems. Still do, actually. Maybe it was just that we were finally dealing with concrete objects, rather than just bunches of "somethings." All that mattered to me was that we were talking about definite objects, not just throwing numbers around.

That is what attracted me to physics, too. Physics is all around us, and whether we were talking about Farmer Frank driving his pickup truck off a cliff or the coefficient of friction between steel and concrete versus that of rubber and concrete, I could visualize what was going on.

In thinking back, I suppose that one reason I liked story problems was that I could use both halves of my brain to figure out what was going on, rather than just the left, logical side.

I remember one from calculus class in which we had to find out whether the Indians who sold Manhattan Island to the Dutch had gotten a good deal or not. Of course, since the island was actually the territory of another tribe (the ones who sold it were just passing through), one could say that they got a good deal no matter how much they were paid.

In any case, we had to compare the current value of Manhattan Island with what the Indians could have gotten, had they invested the $24 worth of trinkets for 400 years at various rates of interest with compounding. If memory serves, they swung a pretty good deal, but that is beside the point.

In that example, what we were really talking about was money, one of my favorite subjects, and one which most people spend time thinking about.

Story problems are an excellent way of taking the skills you have learned in the last five chapters and applying them to your daily life, which is what most peoples' goal is in learning math. They are also an excellent way to teach yourself to "separate the wheat from the chaff." Story problems help you figure out what to concentrate on when you are trying to solve a real-life dilemma.

Story problems are no more difficult than the math puzzles you just got done completing in the workbook. There are a couple of things to keep in mind, but once you understand what is going on and what the puzzle is asking for, they are pretty straightforward.

## *Official procedure for solving story problems*

1) Read the story problem all the way through, then read it again. Decide what it is that the puzzle is asking you to do. What pieces of the puzzle do you have, and which ones are missing?

2) Physically cross out any unnecessary facts. In order to find the length of a fence, for example, you really do not need to know that the owner of the fence, Frank, has neighbors who are named Eric and Susie.

3) Draw a picture, if you can, and label the appropriate parts.

4) Set up the pieces to your puzzle in an appropriate way to provide the answer.

5) Perform the calculations, keeping in mind that your answer will not be simply 4, but "4 oranges," or whatever.

6) Look at the answer to see if it seems reasonable, under the conditions

set down in the puzzle. If 1 orange = 10 apples, can 15 oranges really equal 186 apples? Check your work with check digits, if appropriate.

☆ **Let's try a couple of easy ones:**

Jake and Sue want to start a savings account so they can make the down payment on a house. Each of them is able to deposit $100 per month. How much will they have in a year? In 18 months?

What we are talking about here is simple multiplication. Jake and Sue are each going to deposit $100 per month, so we have a total deposit of $200 per month.

The terms of their deposits are one year (12 months) and 18 months. The only piece of the puzzle we are missing is how much they will have at the end of each of these terms. Thus, if we multiply $200/month by 12 months, we get $2400. For 18 months, we get $3600.

☆ **Here is another one:**

After saving their money in Midas Bank for a year, Jake and Sue found a bank that offers interest on savings accounts. Being canny consumers, they switch banks. Now their money earns 3% simple interest, compounded annually. Assuming that their average balance this year was $3600, how much interest do they earn at the end of the year?

This one is another multiplication puzzle. The parts we have are the interest rate (3%) and the average balance, $3600. To find the amount of interest generated, multiply $3600 by .03, which gives us $108.

Story problems can deal with conversions, too—finding out things like, "How many kilometers are there in 50 miles?" Obviously, before you are able to work a conversion problem, you need to know how many of one thing equal one of another thing. A story problem will generally tell you, and if you are working a problem from real life, you will know (or at least be able to look it up).

☆ **Let's try a conversion problem:**

*Farmer Frank does not believe that banks are safe, not even Midas Bank. But he is no dummy; rather than stuff his mattress with cash, he decides to trade his pumpkin crop for gold, which at least holds its value in the face of inflation. Farmer Frank finds he is able to trade 17 pumpkins for one ounce of gold. If he grows 423 pumpkins this year, how much gold can he trade them for?*

Again, forget Farmer Frank, what he thinks about banks, and that gold holds its value. The pieces of the puzzle are that 17 pumpkins equal 1 ounce of gold, and that we have 423 pumpkins to get rid of. The only thing we are after here is, how many ounces of gold equal 423 pumpkins?

One could, obviously, simply divide 423 by 17, which would leave 24.88 ounces of gold. But what we are really doing when we divide 423 by 17 is setting up the problem this way:

$$\frac{(423 \text{ pumpkins})}{1} \cdot \frac{1 \text{ oz. gold}}{17 \text{ pumpkins}}$$

When we multiply through, just like these were regular fractions (so top-times-top, bottom-times-bottom), we get:

$$\frac{423 \text{ pumpkins} \cdot \text{oz. gold}}{17 \text{ pumpkins}}$$

Since the "pumpkins" on top and the "pumpkins" on bottom are the same term, they cancel out (just like 4/4 = 1), leaving us with:

$$\frac{423 \text{ oz. gold}}{17}$$

Or 24.88 ounces of gold, when we divide it through. While this longer approach may not seem immediately useful, I promise you that it will come in handy when the problems get more complicated, as they always do in real life.

Turn now to the story problems in the puzzle book, and give them a whirl. When you return, we will be talking about how to use math in real life.

# 8 Statistics and probability in real life

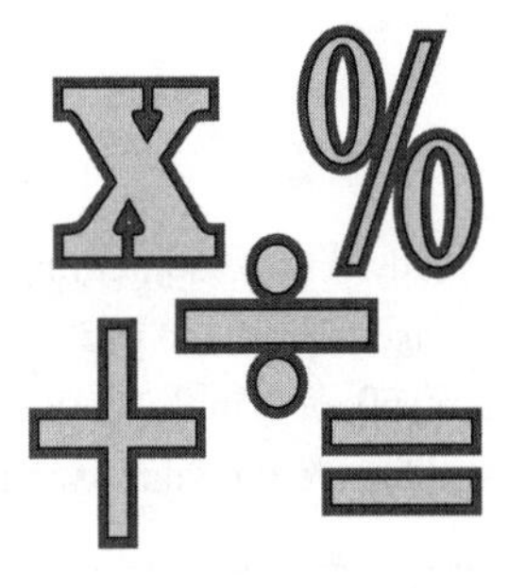

# Chapter 8

## Statistics and probability in real life

While this chapter may not seem to fit in with the others, everyone should have at least a general knowledge of statistics and probability. There will not be much math in this chapter, as a full treatment of statistics and probability would take longer than this book can allow. The best I will be able to do here is try to impart an understanding of how they work, and hope that you get interested on your own. There are some fine introductory books listed in the "Recommended Reading" section at the end of this book.

It is important to understand statistics and probabilities because, like physics, their applications are all around you.

When a commercial says that 37 percent of dentists favor having all of your teeth removed, that is statistics. When you look on the back of a lottery ticket and see, "Odds of winning 1:4.223," that is probability. Aside from those two examples, statistics and probability affect your life in another very real way: insurance.

Every time you take out an insurance policy you are placing a bet. You are betting that you will get into a car accident or your house will burn down, and the insurance company is betting that your car and house will remain in good condition—while they collect your money.

## Probability

The chances are good that you already know something about probability. If we bet $1on the toss of a coin, I take "heads" and you take "tails," you know that your chances of winning are 50-50, or 1 in 2. These odds are also expressed as 1:2 and $\frac{1}{2}$. Your odds of winning are the same as mine; you would expect to come out even in the end.

But how far away is the end? If we tossed that coin ten times, the chances are not at all good (about 25%) that you would have five dollars and I would have five dollars. There are, in fact, 1,024 different ways that series of tosses could end up.

How did I know that there were 1,024 different possibilities for 10 different throws of a coin that could come out 2 ways? I merely multiplied out 210. Think about it: If I throw a coin once, there are only two (or $2^1$) possibilities: Tails or heads. If I toss it twice, there are four (or $2^2$) different possibilities: Tails, heads; tails, tails; heads, heads; heads, tails. And so on.

But even though the chances of getting heads or tails are 1:2 each time, the odds are much greater that we will not be even at the end of ten tosses than that we will be. Why? Because there are many more ways to come up with a non-even result than there are to come up with an even one.

When we say the odds of a certain event occurring are 1:2, that means that after a million, a billion, or a trillion different trials, the numbers will tend to "even out," or approach an equal ratio. I mention this only because the question of odds on things like football teams, lotteries, slot machines, and being hit by lightning come up every once in awhile.

There is one place where you hear probabilities every day: during the weather report. When the weatherman says there is a 30 percent chance of rain the next day, what he means is that since they have been keeping records, on days where conditions were just like they will be tomorrow, it rained 30 out of 100 days.

By the way, I thought I should mention the probably obvious fact that just because the odds on a certain occurrence are 1:2, like our coin, does not mean that a particular event happens every two trials. That is, our chances against getting a series of coin tosses to alternate heads and tails with every toss are 2:1024, or 1:562. As a matter of fact, if you ever get a chance to look at raw data on something like coin tosses, you will likely notice the truth in an informal law of probability: "Stuff happens in clumps." Only rarely will you have strings of data which alternate heads and tails; it is much more likely that you will notice two heads, then three tails, then one head, and so on.

The truth is, the odds on things only describe what we expect to happen, not what actually does. If the odds on a lottery ticket are 1:4, we would expect, if winning tickets were evenly distributed, that over the course of a person's lifetime, he would have won $1 for every $4 he spent on the lottery. Quite the deal.

Anyway, this does not mean that a person cannot spend $1 on a scratch-off ticket and get a $5,000 winner the only time he plays. Obviously, this happens, and the probabilities themselves could not care less. The only result of this kind of luck is to make someone else's lifetime average lower, since lottery tickets are printed in batches with strictly controlled numbers of winners and losers. In a truly random system, however, like a roulette wheel, one man's win does not denote another's loss.

Casinos and insurance companies may have only a 5 percent edge (that is, your odds are 95:100), but they know that in the long run, they will make more money than they give out. If you try to give yourself an edge, perhaps by counting cards at blackjack or burning down your house, you will find that they tend to get rather annoyed.

Probabilities are generally calculated by the following equation:

$$P = \frac{\text{Number of ways desirable outcome can occur}}{\text{Number of all possible outcomes}}$$

Thus, to calculate the odds of rolling a "seven" on two dice, I would first need to look at how many ways I can make a seven. As it turns out, there are 6 different ways that two dice can add up to 7, so I put that number on top of the equation.

Since each die has 6 different sides, that makes the possible combinations 6 • 6 or 36. I put that number on the bottom, so:

Probability of getting a seven: 6 or 6:36
= 36

## Statistics

Figuring out things like the odds of throwing two dice and coming up with a result of 7 is relatively easy. Figuring out the odds of your house burning down or the odds against the Vikings winning the Superbowl is somewhat different. Each of the latter cases require guesswork, or as statisticians like to call it, "margin of error."

Margin of error is usually given as a percentage, say, 5%. When you read "Margin of error ± 5%" in the tiny print at the bottom of a chart in USA Today, what it means is that any percentage given in that chart could be as much as five points higher or lower. They hope the difference is not larger, but that is their best guess.

There is a relationship between probabilities and statistics, in that statisticians use statistics to help them figure out probabilities. That is, they watch what happens in a large number of cases, like cars getting into accidents, and take notes on what sorts of people and cars tend to have accidents.

Males generally have higher car insurance rates than females, for example, because the data which insurance agencies have collected shows more males having accidents. Thus, the insurance companies assign a probability to the chances that you will get into an accident. You, as a factor in the equation, are largely irrelevant. You could be a male who drives very well and still be paying more for car insurance on the same car as a female who drives very poorly.

Your driving record does make some difference, of course; if you have multiple speeding tickets, you will generally pay higher rates, but not because the insurance company is trying to punish you personally. Your rates are higher because they have noticed a correlation between speeding tickets and accidents. Speeding tickets are just another variable, like gender, which increase the odds that you will get into an accident.

## *Correlation is not causation*

Many times, like with smoking and lung cancer, a high correlation means there is some link between two factors; maybe even a causal relationship. Many other times, correlation means nothing. A favorite example of mine is the following:

Since 1945, the number of births per year in Germany has been declining. Also starting in 1945, the population of a particular type of German stork has been declining. Thus, storks really do bring babies.

Obviously this is a somewhat ridiculous example. But I will bet you a dollar that nearly every week when you are watching the news, you will hear of something being correlated with something else, with the implication that the two things are connected. When you hear statements like this, your best reaction is to disbelieve the implied connection.

Good scientists, who (at least in theory) have no personal agenda to push, know that proving a causal relationship is very difficult and time-consuming. Activists and advertisers, who by definition have an agenda to push, will do their best to juggle statistics and, thus, gain attention for their cause or product.

## *The "average" person*

Another thing the media is fond of talking about is the "average" person. If you hear that the average person likes to eat peanuts and you do not, are you abnormal in some way? Should you learn to like peanuts? Do not worry about it. When someone says "the average" anything, they are really not saying a whole lot. Consider the peanut example: If we take a survey of everyone in the country, and ask them to rate their feelings about peanuts on a scale of 1 to 10 (if they answer 1, they hate peanuts, and if they answer 10, they love them), their responses might be much like this:

1,10,1,10,10,10,1,1,10,1,1,10.

If we add all these numbers together, then divide the answer by the number of responses (which is how we get an average), we end up with 66 ÷ 12, or an average of 5.5. Something in the middle of a scale ranging from hate to love might well be defined as "like," but you will notice that not one of the people surveyed rated peanuts anywhere near a 5 on their preference scale. Thus, while the numbers might dictate that the "average" person "likes" peanuts, the truth is that people are strongly divided on peanuts—they either love them or they hate them.

This particular survey could be said to have a large "standard of deviation." The standard of deviation is simply a tool used to convey how closely packed the numbers are. In the peanut example, the numbers were not close in agreement at all. But in another case, they might be.

Say we would like to buy some 12-foot boards for a fence that we are building. We go to the lumber yard and are told that two different stacks of boards each contain boards which average about 12 feet. If an average length were all we needed, we could obviously take either pile, but since we are canny consumers, we decide to examine each of them.

After looking at all the boards in the first pile, we determine that it contains two types of boards. Some are 6 inches long, while others are $23\frac{1}{2}$ feet long. Obviously, the 6-inch boards will not do us much good, and neither will the 23½ footers, as we do not have a saw (we are better consumers than we are carpenters). This pile will not do, as the standard deviation is too large; we decide to examine the second.

The second pile also contains two types of boards—half are 12 feet, 1 inch long, and half are 11 feet, 11 inches long. This pile has quite a small standard deviation, and we decide that we can make the different sizes work, so we buy them.

Another factor to consider when trying to evaluate the truthfulness of statistics is sampling size. In the above example, we looked at every board in both piles. While measuring every part of a group is the best way to ensure accuracy in our figures, it is rarely practical.

The U.S. census, for example, costs millions of dollars and takes a whole year to try to get some figures on the American population, but it is still off by 5 or 10 percent. Most people or corporations who need information about a particular group of people or things simply cannot go to the time and expense of taking a measurement on every item or person in the group. They usually have to take a sample of the group and try to make educated guesses about the whole group from what they find out about the sample.

But just taking a sample introduces quite a few ways to make an educated guess less accurate. The drug Prozac, for example, was tested on ridiculously few people before being introduced to millions in the American market. Because their sample size was so low, the company which released Prozac found out a lot of new things about their drug after it had been prescribed to millions of people—and not all of what they found out was good.

A mistake like this can cost tens of millions of dollars in litigation expenses, which is not "good business," as the companies which manufactured Thalidomide and the Dalkon Shield can attest. Had the company used a larger sample size, they might have avoided or been better prepared for this litigation.

What size is large enough to be a "good" sample size? It depends on the situation. Generally, the larger the better, as time and money allow.

Another thing to keep in mind about sampling is where the sample is drawn from. If I wanted to "prove" that most people believed in space aliens I'd release a study which said so, and would most likely draw my sample from those known to feel this way. Is it a lie to engineer a study in this way? Well, perhaps not technically, but it certainly is junk science.

Surveys are a popular way of determining popular opinion in today's world, but they are perhaps more prone to error than any other sort of measurement.

First, it is generally only the people who feel strongly about an issue who respond to surveys. How often have you seen news programs give two different phone lines (one "pro," one "con") to call in order to get a sampling of public opinion? Such surveys, however, are nothing more than junk. There is nothing to prevent a person from calling in several times to register his vote, should he feel strongly one way or another. And, like I said, the only people likely to call are those who feel strongly.

Second, people lie. They lie for a great number of reasons, many of which are not known to statisticians, but they lie just the same. If a person just finishes voting on a controversial issue, for example, and she is pulled aside by a television reporter who asks her how she voted, she may respond in a way that her friends and relatives will approve of, rather than with the truth.

The above example brings up an important point. People can and will change their answers to questions depending on who is doing the asking. Had the above-mentioned reporter asked our voter off-camera how she voted, she may well have answered a different way. Gender and ethnicity differences between surveyor and respondent have also shown to make quite a difference. Some whites, for example, may feel uncomfortable telling a black researcher that they feel affirmative action programs have gone too far, while they may quickly tell a white researcher more than he cared to know.

## *Charts and graphs*

Another way in which data is consciously manipulated is to goof around with charts and graphs. A couple of years ago, my local paper showed a graph which purported to show the decrease in traffic fatalities due to laws requiring seatbelt usage.

The data on the chart went back to 1973, and there were two significant "dips" in the line measuring fatalities. One was in about 1976, when the nationwide 55 mph speed limit was established, and the other one was in the late 1980s or early 1990s, when many states passed mandatory seatbelt usage laws. The story which accompanied the graph, however, tried to claim that the total decrease in deaths since 1993 was due to increased seatbelt usage. What about the lower speed limit? What about tougher drunk driving laws? What about anti-lock brakes and airbags?

When reading supposedly factual statistical information, always think to yourself, "Is there another way that the data could be interpreted?" Usually, there is.

There are a few different aspects to every decent line graph. Every graph must have a "zero point"—the place where the horizontal axis ("*x* axis") and the vertical axis ("*y* axis") intersect must be labeled "0."

Second, the markings on both the x and y axis must be evenly separated and marked. That is, the y axis can be marked "0, 100, 200, 300," and so on, but it cannot be marked "0, 100, 500, 1000," as the units are not evenly spaced. Units which are not spaced evenly make the graph virtually useless.

Third, the *y* axis must be about three-quarters as high as the *x* axis is long. I do not know who set this standard, but graphs which do not live up to it will be viewed skeptically by anyone who knows anything about statistics. Look at the graphs following:

### *Graph I*

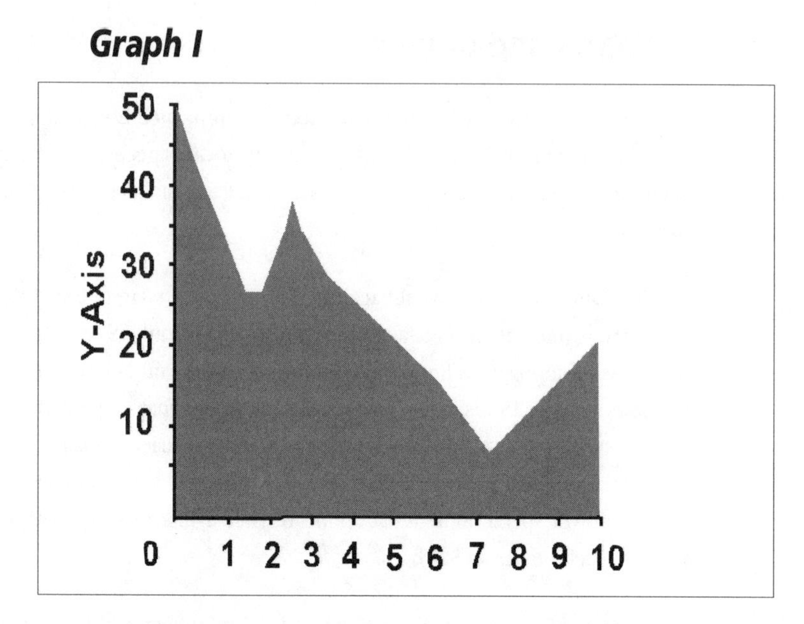


### *Graph II*

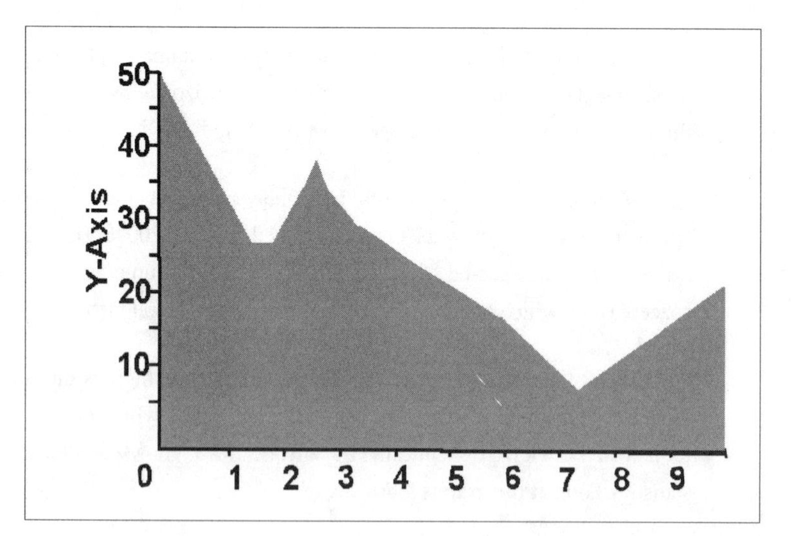

Graph I has a $y$ axis which is as high as the $x$ axis is long; thus, the significance of the data it is trying to show has been exaggerated. Graph II is correctly proportioned. Note that both have a zero point, and the units on their axes are evenly spaced.

Just like line graphs, bar charts, or "histographs" (they are not technically the same thing, but for our purposes, they come pretty close) can be manipulated to show greater differences than actually exist.

Graphs III and IV both show the same data, but the scale on Graph III has been chosen to make the differences in the data seem much more significant than it actually is. Graph IV is somewhat more reserved.

### *Graph III*

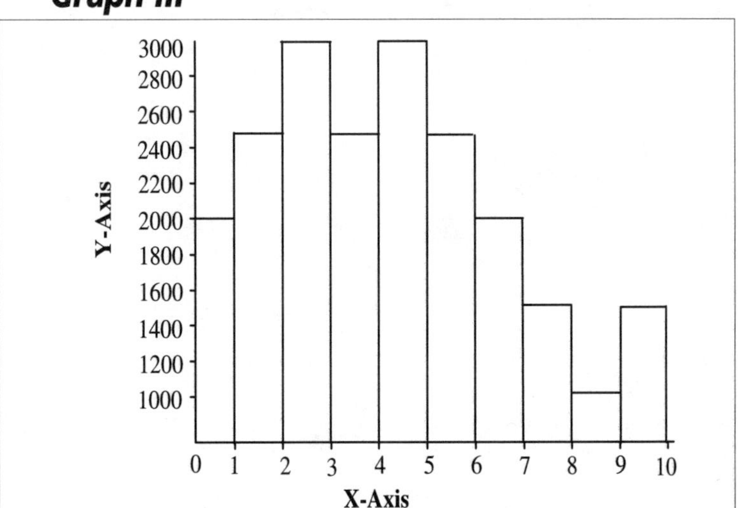

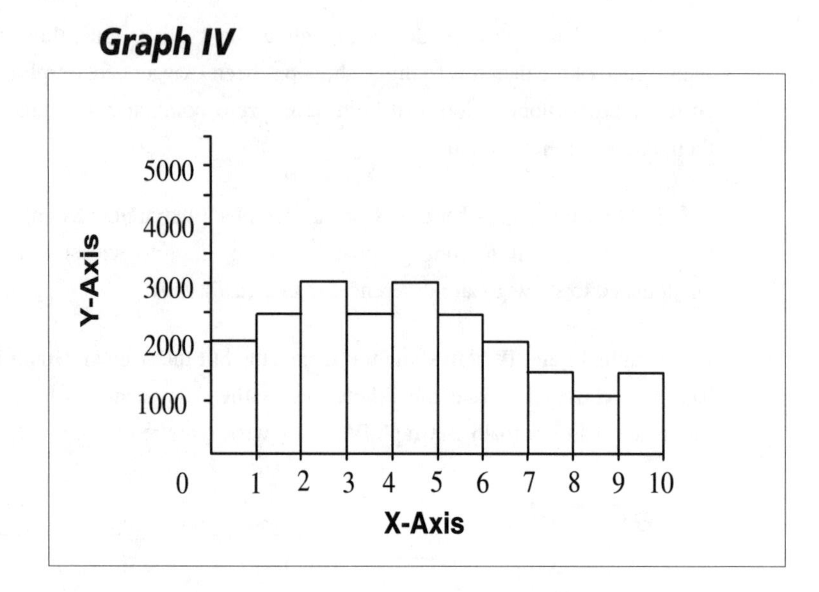
Graph IV
5000
4000
3000
2000
1000
Y-Axis
0
1
2
3
4
5
6
7
8
9
10
X-Axis

Everyday
math
9

# Chapter 9

## Everyday math

As I have said before, you can use your new math skills every day. Figuring out square footage may be useful only occasionally, but determining how much to tip a waiter or waitress may be useful every day.

### *Math on a budget*

From time to time, all of us are short on cash. We may have plenty of money in the bank, but I am referring to the times when we need to stop by the store just for a couple of things, and there is not a convenient ATM nearby. In these cases, it is important to figure your grocery bill accurately—there are few things which embarrass me more than being short of money at the cash register with a line of people behind me. Not having enough money to pay for the dinner you have just eaten can be even worse.

In these cases it is important to overestimate your bill. With the skills you have accumulated, the calculations should be a snap.

When grocery shopping, I generally keep a running total in my head, rounding each new item up to the nearest quarter. But I am never sure which items have tax and which ones do not. I also have a sneaking suspicion that taxable and non-taxable items sometimes vary from store to store, depending on what the clerk has been told. In order to be safe, I assume that tax will be charged on each item, and I keep a separate tally in my head for that. Here is the formula for a grocery bill:

$$(P1 + P2 + P3...) + [T \cdot (P1 + P2 + P3...)]$$

P1, P2, and so on are the prices of the individual items, each rounded up to the nearest quarter (to reduce addition mistakes), and T stands for the tax rate where you live. The brackets, or [ ], function just like parentheses do, and as the order of operations tells us, you work from inside the parentheses to inside the brackets, to outside the brackets.

Below are a list of items that I typically purchase, along with some prices which may or may not have a basis in reality:

| Item | Actual Cost | Rounded Cost |
| --- | --- | --- |
| Coffee | $ 7.23 | $ 7.25 |
| Paper Towels | $ 1.19 | $ 1.25 |
| Milk | $ 2.09 | $ 2.25 |
| Hamburger | $ 2.79 | $ 3.00 |
| Totals: | $ 13.30 | $ 13.75 |

I then multiply my rounded total of $13.75 by .065 (6.5 percent), bearing in mind that I can fudge a little because all of my prices are rounded up. It is a lot easier to multiply by .05 than it is by .065, so I will do that, trusting that I will still have enough money to cover the tab.

13.75 • .05 = .6875, or $.69 tax, making a total of $14.44.

Which, considering that some of the above items should not have tax charged on them, should be close enough. How close were we to a dead-on total? As I understand the sales tax laws, only the paper towels should have had sales tax charged. This means that the formula to figure out the exact total should look like this:

(13.30) + (1.19 • .065) = total

When calculated, it gives us a total of $13.38, or almost a dollar less than our estimate. Remember, that if you are shopping in a different state you may have no idea which items are taxed and at what rate. If this is the case, you may wish to round to the nearest 50 cents, rather than the nearest quarter.

Adding up a restaurant tab is much the same as adding up a grocery list in theory. The writer Douglas Adams mentions in one of his books that mathematical calculations on restaurant tickets work differently than anywhere else in the universe, and it seems to be true.

It seems like every time I go to a restaurant, I order an entree which costs something like $6.99 with a glass of beer for $2.75, and my sweetheart does the same. When the check comes, however, it always seems to be about $28.

Figuring out the bill in a restaurant when you have limited cash differs in two ways from adding up a grocery bill: 1) unless you wish to be thought of as a cheapskate, you must leave a tip of some sort, and 2) if you miscalculate, you end up washing dishes for an hour.

Other than that, the calculations are pretty much the same. Here is the equation:

$$\{(P1 + P2 + P3...) + [T \cdot (P1 + P2 + P3...)]\} + (G)$$

P1 and so on are the prices of items you wish to order, T is tax, and G is gratuity, at whatever percentage you think the waiter deserves. Notice that in many states absolutely everything you order in a restaurant is taxed, and also notice that the gratuity you leave is calculated on the total price of the food plus the tax. As with the grocery bill, I recommend that you round each food or drink item up to the nearest quarter. As restaurants often list their prices as $6.95, 4.95, and 7.95, however, you may wish to inquire about the tax rate before you order (only if your estimated tab comes very close to the amount of cash you have on hand) because you are only giving yourself a nickel leeway with each item added to your tab.

Another interesting (or, sometimes, depressing) way to use math is to figure out how much, exactly, you earn per hour. While you may have been told one figure when you were hired, that figure is in pretax dollars, so your net earnings per hour will be lower.

Payroll departments can and do make mistakes when figuring out your paycheck, and if you are on salary, you will undoubtedly notice the difference. But hourly workers' paychecks often vary from week to week, depending on how many hours they put in.

To find out your hourly rate, simply take the amount of your check and divide it by the number of hours you worked during the current pay period. Do this two or three weeks in a row (improving your sampling size to make sure the first check was not in error), and you should have a firm figure.

For every paycheck in the future, then, you can multiply the number of hours worked by your net hourly rate to see if your result equals the amount of your paycheck—at least until you get a raise, at which point you will have to figure the whole amount over again. If your figures do not agree with what is written on your paycheck, check your work using check digits. If your figures check out, the payroll department has made a mistake.

Another time when "math on the fly" is useful is when you happen to see a sale at a favorite store. Say you had your eye on a CD player which normally retails for $300. This week, however, Larry's Loose Wires is having a blowout sale—everything in the store is 25% off. How much does the CD player cost now?

There are two ways to approach a problem like this: 1) you could multiply $300 by .75, as 100%–25% = 75%, or 2) you could multiply $300 by .25, then subtract that answer from $300. Either way you get the same answer: $225.

## *Non-monetary calculations*

Many people like to check the mileage they are getting with their cars; they may use the figures as an indication that they need to get a tune-up, or to see whether they get better gas mileage with a higher-octane blend of gasoline, or other reasons. To figure out your gas mileage, simply record your car's odometer reading when you fill the tank. Then next time you fill the tank, take note of the new odometer reading, and subtract the old from the new. If you have a trip meter on your car, simply reset it each time you fill up. Then divide the total number of miles you put on since the last fill by the number of gallons (not the dollar amount) it took to fill your tank this time. The equation looks like this:

$$(M_{new} - M_{old}) \div G$$

Where M stands for miles on the odometer and G, in this case, stands for gallons. The letters we use for variables really are not important; it is the concept which is important.

***Note***: This method will not work if, instead of filling the tank, you simply put in $10 worth of gas, because you will not get an accurate number of gallons by which to divide your mileage.

The next couple of techniques may not be useful every day, but they will be useful eventually.

Figuring out perimeters: "Perimeter" is just a fancy way of saying "the distance around something."

Say you wish to put up a chain-link fence around some property you own. You need to know how much fence to order, but you do not want to order too much, as it would be a waste of money. It is a fairly large piece of property, so measuring the perimeter with a yardstick would take a long time. You would rather spend twenty minutes doing a little measuring and three minutes doing a little math than an hour and a half measuring.

If you had a perfectly square piece of property, you could just measure one side and multiply that result by

4 (L • 4). Unfortunately, few pieces of real estate are perfectly square.

If your property was a perfect rectangle or any other shape that had two sets of parallel sides, you could measure two adjoining sides and multiply by two ($L + W \cdot 2$).

If the property was a right triangle, you could measure the two short sides length and height, square each of them, add them together, and take the square root of that figure to get the length for the diagonal side ($L^2 + H^2 = D^2$), then just add them together.

If the property was a circle, you could measure the distance across (the diameter), divide that by 2 to get the radius, and multiply that number by $2\pi$ to get the perimeter. $\pi$ (or "pi") is roughly equal to 3.14159.

You may not be itching to put up a fence, but those are the equations for finding the perimeters of many common shapes, and they can be very handy to have available. If you need to buy some wallpaper to run a border around the ceiling in a room, for example, you will need to figure out the perimeter of a rectangle. If you cut a 4 foot by 8 foot sheet of plywood diagonally and do not have a tape measure handy, you can now figure out the length of the remaining side ($4 \cdot 4 + 8 \cdot 8 = D^2$, or a hair shy of 9 feet).

Figuring out areas is just as easy:

☆ To find the area of a square, measure one side and square the measurement.

☆ To find the area of a rectangle, multiply two adjoining sides.

☆ To find the area of a right triangle, multiply the two short sides and divide by 2.

☆ To find the area of a circle, multiply the radius by itself, then by $\pi$.

What if you need to find out the amount of square feet of your living room walls so you buy enough paint, but your living room is irregularly shaped? No problem. The chances are quite good that even though your living room is irregularly shaped, it is still made up of rectangles. If your living room is wide on each end, but the middle is narrow, measure each stretch of flat wall space separately. Each side of a corner is simply another rectangle, after all—just add them together.

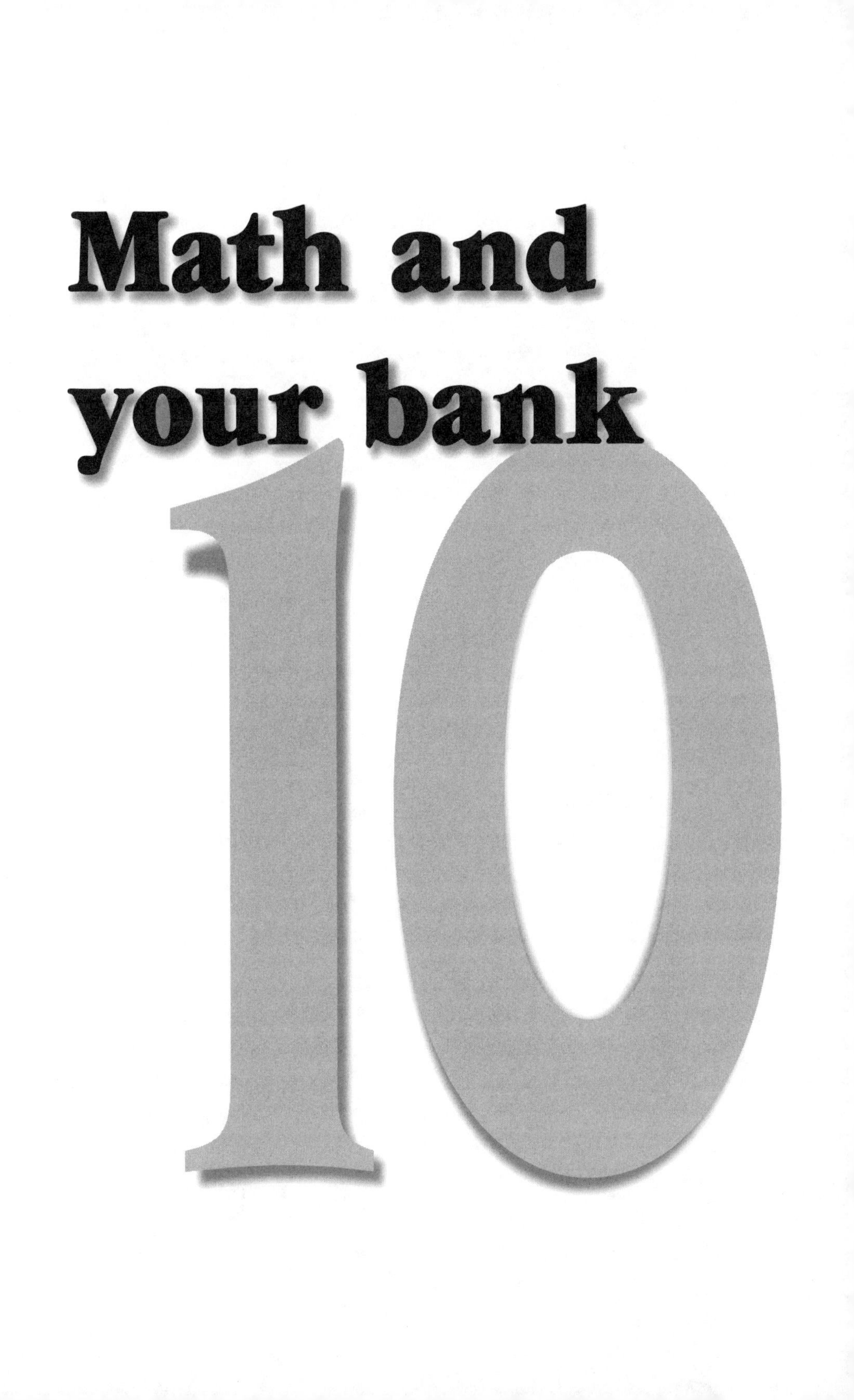
Math and
your bank
10

# Chapter 10

## Math and your bank

### *Warning!*

Many of the calculations involved with this chapter will require a calculator with lots of buttons. Actually, the only button you need to look for is one that says "$xy$" or similar. We will be raising numbers to the 360th power, and I do not want you to stay up all night hitting the "$x^2$" button repeatedly.

This chapter actually involves you and your money in all sorts of official transactions. Sooner or later, everyone has enough money to worry about, and most of us keep it in a bank. Once in awhile, we get worried because we do not have enough money, and we go to banks then, too, for a loan.

Yet many of us do not shop around for the best deal—we have one bank, and we go to that one when we want to save or borrow some money. But in the same way that you would shop around for a new car to get the best price, it is also important to shop around for the best bank rates.

In this chapter, we will not only cover math, but also we will apply what we have learned in the past few chapters on how to be a smart consumer when dealing with banks, investments, and finance companies.

## Interest

We learned how to calculate simple interest by multiplying the rate, or percent, by the amount our fictional characters were saving. Most savings accounts and CDs, however, do not offer simple interest, but compound interest.

Compound interest is one of the great inventions of the 20th century. When a lending or savings institution calculates compound interest, what they do is take a look at your average balance over a particular period (a day, a month, or, more commonly, three months), calculate the interest you have earned in that period, then add it to your account or CD. It really adds up when dealing with larger sums, and the longer you leave your money in their institution, the better the deal.

**Compound interest is calculated using the following equation:**

$$T = P(1 + I)^n$$

The variables are defined in this manner:

☆ T is the total amount of money we end up with, or the principal plus accumulated interest.

☆ P is the principal, the amount of money we start with.

☆ I is the interest rate we get, divided by 100 because we are translating from a percent to a decimal. We also divide the stated rate by the number of times per year the interest will be compounded—for quarterly compounding, we would divide by 4, for example.

☆ n is the number of interest periods that we leave the money in the bank. If our deposit was compounded quarterly, for example, and we left the money in for a year, n would be 4.

Let's try a couple of examples to show the difference compounding can make:

You want to invest $2,000 for two years. Your decision has come down to two different banks. One offers 3.9 percent simple interest annually, the other offers 3.9 percent compounded quarterly.

At the first bank, your account balance (principal + interest = earnings) after two years would be $2,000 + (2,000 • .039 • 2) = $2,156. We multiplied the interest accrued by 2, remember, because you left it in for two years.

At the second bank, your interest earnings would be $2,000 $(1 + .00975)^8$, or to simplify a little bit, $2,000 (1.0807), or $2,161.40

What? Only $6 difference? Three dollars a year is not that big of a deal? Okay, you are right. But what if we could find a bank where they compounded monthly, rather than quarterly? Then we would have $2,000 $(1+.00325)^{24}$, or, after simplifying a bit, $2,000 (1.081), or 2,162. Hmm, only .60 higher.

Let us try a different bank, one which compounds interest daily:

$$\$2,000\ (1 + .000106)^{365}$$

That comes out to be $2,078.89, or a difference of $22.89, or $11.50 (roughly) per year for being a careful shopper. I know it does not look very impressive, but that is because interest rates are low in the example.

Let us try one with a higher interest rate. Suppose you got an offer to loan the bank money at 18% per year, compounded monthly. You can only afford to loan out $1,000. Here is how it looks on paper:

$$T = 1,000\ (1 + .015)^{12}$$

When solved, this comes out to be $1,195.62, or $15.62 more than you would have gotten at simple interest. What a deal! I will bet you wish this deal actually existed. Well, it does, except rather than you loaning the bank money by depositing it, the bank is loaning you money —this is the interest you may be paying on your credit cards. Every time you charge something on a credit card, you are writing yourself a loan at 18%. Think about that next time you just have to buy a particular item.

If you have $2,000 in your savings account, and you have an outstanding credit card bill of $1,000, you will save a fair chunk of money by taking money out of your savings to pay off your credit card completely. Let's look at an example:

You have $2,000 in savings, earning 3.5 percent interest, compounded quarterly. You also have a credit card with a $1,000 balance. The rate on the card is 18 percent, and because they add interest to your balance each month, we know it is compounded monthly.

If you left your money in savings for a year, you would earn $T = 1{,}000\ (1 + .00875)^4$, or $1,035.46—you would earn $35.46 in interest.

Assuming an average $1,000 balance on your credit card, however, by the end of the year it will cost you $T = 1{,}000\ (1 + .015)^{12}$, for a year-end balance of $1195.62, or a loss of 195.62 in interest. By leaving the money in savings, you are losing $160.16.

Do you want the credit card which offers 18% interest and no annual fee, or the one that charges 12% with a $50 annual fee? Instinctively, the 12% one sounds better, but how much better, exactly? How many hours are you going to have to work to pay the interest charges, given an average balance of $750?

Well, shall we figure it out? The first card charges 18%, compounded monthly, with no annual fee. We are assuming an average balance of $750. Thus, we would set up our equation like this: $T = 750\ (1 + .015)^{12}$. Doing the math tells us that this card would cost us $146.71 per year (896.71 – 750).

The second card only charges 12%, compounded monthly, but it has a $50 annual fee, so we need to add $50 to whatever the interest is going to cost us. Thus, the puzzle will look like this:

$T = 750\ (1 + .01)^{12} + 50$, which, when solved, gives us a total cost of $145.12 (895.12 – 750). Pretty darn close, but the second card is still a little cheaper.

## Mortgages

Mortgages are also a good way to make sure your bank shows a healthy profit. If you find your dream house for $150,000, how much will your payments be? The loan amount will actually be closer to $135,000, as you usually have to put up 10 percent or so as a down payment. Here is the formula for calculating mortgage payments:

$$P = \frac{(\text{Amount of Loan}) \cdot I\ (1 + I)^n}{(1 + I)^n - 1}$$

In this case, "P" stands for monthly payment, "I" for the interest rate in decimal form, and "n" for the number of payments in the loan.

30-year mortgages seem to be all the rage nowadays, so we will calculate that first. We will use 10% as the interest rate. On installment loans, like mortgages and auto loans, the quoted rate is compound, rather than simple interest. In this case, "n" is equal to 30 • 12, or 360, and "I" is .10/12 or .0083. As it looks on paper, then:

$$\frac{(135{,}000) \cdot .0083\ (1 + .0083)^{360}}{(1 + .0083)^{360}}$$

Looks pretty impressive, but it is really not all that scary. After the first step, the puzzle looks like:

$$\frac{(135{,}000) \cdot .0083\ (19.6027)}{18.6027}$$

The next step looks like:

$$(135{,}000) \cdot .0087461$$

Finally we come up with the answer of $1180.73 as a monthly payment.

Now we come to the fun part: How much is your $150,000 house going to cost you? You have already paid $15,000, so we will add that in when we figure out the total of your mortgage payments. You will be paying $1180.73 every month for 360 months. Multiply those two together, and add the $15,000 down payment, and we come up with a grand total of $440,064.24. Wow! Almost half a million dollars for a $150,000 home!

How can we cut that amount down? One way is to reduce the length of the loan, paying more each month. Intuitively one would think that paying a loan off in $\frac{2}{3}$ the time would mean that the payments would have to be $\frac{3}{2}$ of what they are now. Not so. Paying off this loan in 20 years, rather than 30, would call for monthly payments of only $1309.55 and would bring the cost of your house down to $329,292.52. So by paying an extra $129, roughly, per month, you can save $110,000 over 20 years. Pretty good. Now let us see what a ten year mortgage would cost:

A ten year mortgage would have payments of $1,781.01, and your dream house would cost you $228,720.86. Of course, getting a lower interest rate or putting up more of a down payment would further reduce the house's actual cost. Those options would require some shopping around, however.

Your bank loves to give out installment loans, like mortgages and auto loans, because they earn compound interest. As you can see, when we set compound interest to work on large amounts and long terms, it really adds up in a hurry.

When shopping around for a car loan, it is in your best interest to turn down installment loans for just this reason. The price of some new cars, after all, approaches the price (and sometimes exceeds the price) of a small house. Instead, tell (not ask) the banker that you want a simple interest, single payment note that allows for monthly payments. Your banker will tell you that the bank does not do that. He lies. If necessary, shop around until you find a banker who will write you a simple interest loan.

In addition, I recommend never purchasing the "credit life and disability" option on any loan, and that includes credit cards. There are three reasons for this: One, the insurance is generally overpriced, and two, the purchase price of the insurance is stacked on top of your loan, so you are paying interest on that too! Three, the insurance you purchase only covers the loan itself, not any other loans you have taken out. If you pay off all but $100 of your loan and then become disabled, the insurance pays only the remaining $100—no more.

If you feel that you are likely to become maimed or die during the course of your loan, see your insurance agent. He will offer you cheaper coverage without extra interest charges.

Luckily, compound interest can work for you too, through instruments like CDs, IRAs, and 401(k)s. Actually, you could say that paying your credit card balance in full every month is just like "earning" 18 percent compound interest, as you are saving yourself money which you would otherwise be paying out.

If you are stuck trying to decide between two different investment deals which have different rates of interest and compounding, there is a way to figure out the effective annual yield of each. "If this compound rate were simple interest, what rate would it be?" Why bother? Because it lets you compare daily versus quarterly compounding with less math than working out the compound interest formula.

Here is the formula: $1000 (1 + I)^k - 1000$, where you know our old friend "I," the decimal form of the quoted rate divided by the number of times per year interest is compounded, which "k" just happens to represent. This formula calculates the amount of interest generated at the effective rate if $1,000 is left in the account for a year. Time for an example:

Harvey is looking at two different CDs. One offers weekly compounding at 5.9%, while another offers daily compounding at 5.7%. Find the effective annual yield for each.

For the first CD, the formula looks like this: $1{,}000\,(1 + .0011346)^{52} - 1{,}000$. When solved, we come up with an answer of $60.74 earned per year. Slide the decimal point one space to the left, and we get the effective annual yield for this CD, which is 6.07%.

The second CD looks like this: $1{,}000\,(1 + .0001561)^{365} - 1{,}000$. When we work through the puzzle, we get a result of $58.65 in interest, or 5.87% effective annual yield. Obviously, the first CD is the better deal.

## Other investments

Many other investments are not as easy to plan as saving money at a bank. Real estate, stocks and mutual funds, rare coins, and bonds may all lose money or they may have great performance. You cannot pre-determine the rates of return on things like this; all you can do is look at how they have performed over the last 10 or 20 years.

You can, of course, review your investments from time to time and see how well they have performed since you bought them. This is a rather easy process. The formula is:

$$\%\ \text{Increase} = (P_1 - P_0) \div P_0 \cdot 100$$

$P_0$ is the original purchase price, $P_1$ is the current value. If you had bought stock in the Whoopee Pharmaceutical Company, for example, at $15 per share, and it is now worth $18 per share. The percent increase is $(18 - 15) \div 15 \times 100$, or 20%. Easy, no?

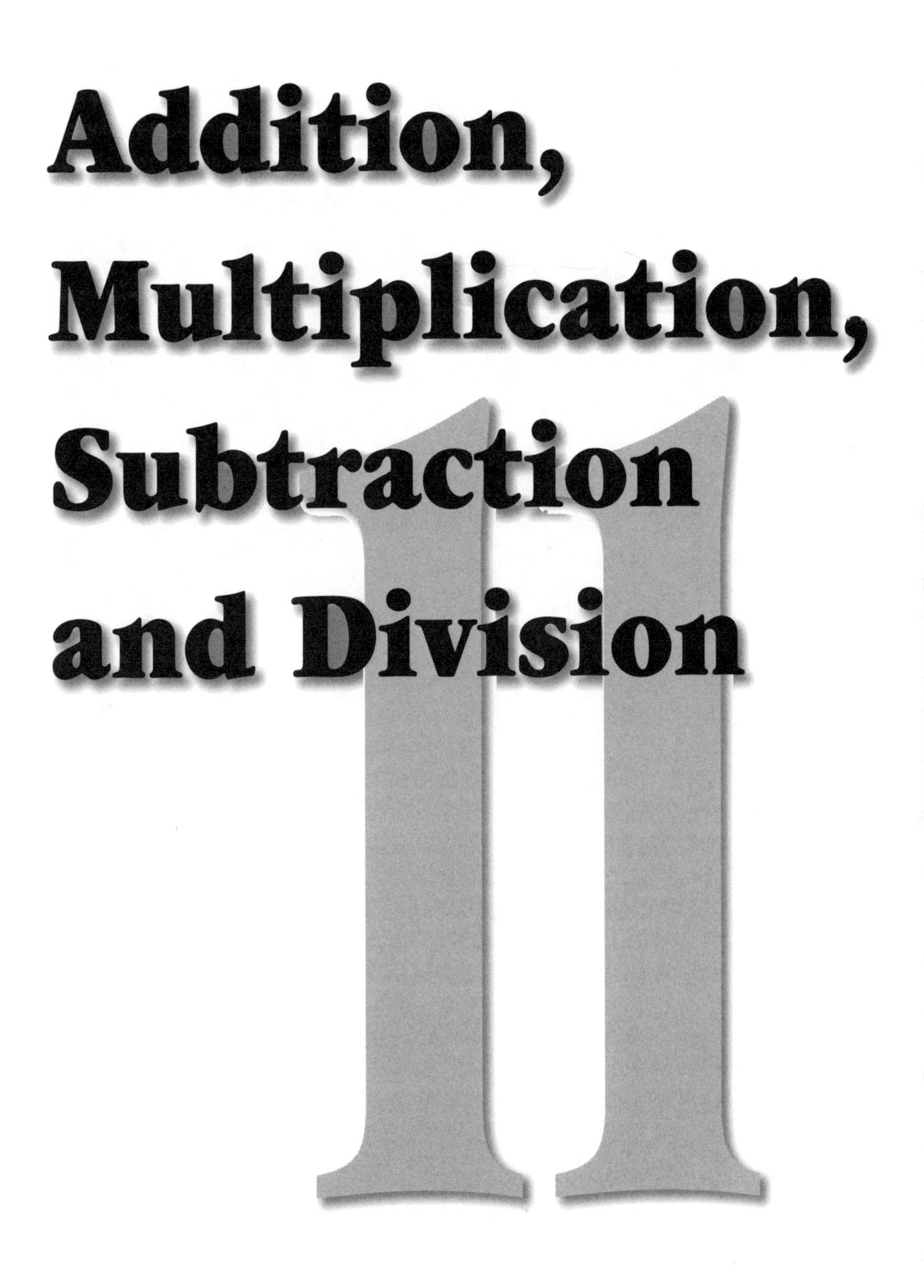
Addition,
Multiplication,
Subtraction
and Division
11

# Chapter 11

## Addition, Multiplication, Subtraction and Division

## Addition

Round whenever possible. Generally the mathematical standard for rounding is that 1, 2, 3 and 4 get rounded down to 0, and 5, 6, 7, 8 and 9 get rounded up to 10. Decide how accurate you need to be, which may depend on the accuracy of the information you are working with.

Numbers can be added in any order, the answer is always the same.

Make tens to add columns of numbers quickly. Go up and down the column and cancel out any numbers you can add together to make a ten. Keep track of tens separately, and add them to whatever numbers are left.

```
 7    7
 4    4
 3    3
 8    8
 9    9
 4    4
 2    2
 5    5
+7   +7
```

In the previous example, "tens" are underlined. The rest of the numbers may be added normally, taking 10s from the running total and adding them to the 10s pile.

Remember to pick numbers apart to make them easier to work with. 236 is the same as saying "200 + 30 + 6," and so on.

Keep your columns lined up. This is generally done by lining up the "ones" columns of all the numbers you're going to be adding, but there's no shame in tacking on zeroes to the front of a string of numbers if you feel more comfortable aligning them on the left.

When adding decimals, like money, line up all the decimal points, rather than using the ones column.

If it helps, round out the numbers by adding or subtracting the same number from both the top and the bottom numbers, then add or subtract double the number from your answer. This method works well for addition and subtraction, but not multiplication or division.

Always multiply left to right—it is faster, and every step brings you closer to the answer.

Check your work by finding the check digit for each row, then add them and compare them with the check digit of the answer, like so:

```
  1234    1
+ 2110    4
  3344    5
```

# Multiplication

Multiplication is really just a shorthand method for adding. When I say "four times five equals twenty," what I really mean is "take four sets of five somethings, add them up, and you get twenty." In mathematical parlance,

$$4 \times 5 = 5 + 5 + 5 + 5.$$

When multiplying larger numbers, the first thing you should do is factor the two numbers, because it is much easier to multiply a number by a one-digit number than a two- or three-digit number.

## *Rules for factoring*

A number is divisible by 2 if it is even; that is, if it ends in 2, 4, 6, 8, or 0.

A number is divisible by 3 if the number's digits, added together, equal 3 or a multiple of 3.

Example: Is 27 divisible by 3? Yes, because 2 + 7 = 9, a multiple of 3. Is 127? No, because 1 + 2 + 7 = 10, which doesn't divide evenly by 3.

A number is divisible by 4 if it is even and its last two digits are evenly divisible by 4. 1,632 is divisible by 4 because both 1,600 and 32 are divisible by 4.

A number is divisible by 5 if its last digit is a 5 or a 0.

A number is divisible by 6 if it is divisible by both 2 and 3.

Remember to factor completely.

If splitting numbers up as factors, take the two biggest factors and multiply them together first. Then multiply that result by every remaining factor, in descending order of size.

If splitting up numbers by subtraction (i.e. 63 – 3 = 60) remember that you have to multiply both parts by any remaining factors.

Multiplying any number by 10 is easy—just add a zero to the original number.

You can multiply numbers by 100, 200, and 300 in the same way. Just add two 0s to the end of the answer, instead of one.

Multiplying a number by 5 is just like doing it by ten, except after adding the final 0, you divide the answer by 2.

To multiply by 15, multiply first by 10 and then add half of it to the original answer.

Since there are four 25s in 100, you can always multiply numbers by 25 by first multiplying them by 100 (or tacking on two 0s), then dividing by 4.

Or, multiply by 10, figure out what half of that is, then double the original answer and add the half.

Multiplying numbers by 50 is even easier. Just multiply by 100, then cut the answer in half. 440 x 50 becomes first 44,000, then 22,000, when you cut it in half.

Whenever you multiply a number by 11, 22, 33, or any number that has both digits the same, you can simply multiply the number by 10, 20, 30, or whatever, then add 10 percent.

When multiplying decimals, just treat them like regular numbers until you get your answer. Then, count over, right to left, on your answer and put down a decimal point where you have counted the same number of places as you have decimal places in the whole problem.

When, and if, you come across an exponent, remember that $3^5 = 3 \times 3 \times 3 \times 3 \times 3$, or 243. In the case of $10^6$, you could multiply 10 by itself 6 times, or you could just write down a 1 and add six 0s.

Numbers in science are often expressed as $1.234 \times 10^6$, rather than 1,234,000. If you come up against a decimal like that, rather than add six 0s, you just move the decimal point to the right six places, adding 0s when you run out of numbers.

If you should happen to come across a negative exponent, like $14^{-23}$, all the negative sign means is that you need to move the decimal point to the left 23 places, rather than to the right.

Any number, n, with an exponent of 1 has a value of n, because the exponent isn't high enough to provide n with copies of itself to multiply itself with.

Any number, n, with an exponent of 0 has a value of 1, except 0, which is always 0.

Checking your work with multiplication is much the same as with addition, except you multiply the check digits from the two numbers instead of adding them. Then compare them against the check digit of your answer.

# Subtraction

To subtract any number from any other number, *just figure out what number, added to the one being subtracted, makes the original number.* Using mathematical symbols, then, subtraction is less a question of 234 – 102 = ? than of 102 + ? = 234.

To speed up subtraction, we can steal a few tricks from addition—we can certainly round the numbers, and we can also speed things up by working left-to-right.

We can also add and subtract from the numbers being subtracted in order to make them easier to work with. In the example of 234 – 102, we might subtract 2 from 102 in order to simplify things as long as we remember to subtract 2 from the other number. Otherwise, the answer would be 2 too high.

Subtract from left-to-right, just like we add. When you get to a place where the digit which is being subtracted is larger than the one it is being taken from, pretend the number on top is actually 10 more than it is, and subtract normally. Write the result down, then make a slash across the number before it. It works like this:

$$\begin{array}{r} 4288621 \\ -\,1194203 \end{array} \qquad\qquad \begin{array}{r} 4288621 \\ -\,\underline{1194203} \\ 3\not{1}9442\not{2}8 \end{array}$$

When you look at your answer, "see" each number with a slash through it as one less.

To check your work with subtraction, you add the check digit of the answer to the check digit of the number being subtracted to see if it adds up to the check digit of the number being subtracted from. Example:

$$\begin{array}{r} 100 \\ \underline{47} \\ 53 \end{array} \quad \begin{array}{r} 100 \\ \underline{47} \\ 53 \end{array} \quad \begin{array}{r} 1 \\ 2 \\ 8 \end{array}$$

Since 8 and 2 make 1, this puzzle checks out.

Negative numbers are what you get if you subtract a larger number from a smaller one. If a negative number and a positive number of the same value are added together, they'll add up to 0.

Subtracting negative numbers is a little trickier. When you subtract a negative, you're really adding a positive.

$13 - (-12) = 25.$

Multiplying and dividing negative numbers is not too tough, either. Just remember that in multiplication and division, *negative numbers poison everything they touch*. That is, whenever you multiply or divide by a negative number, your final answer will be affected. When you need to work a puzzle like 45 x - 4, your final answer will be a negative number, –180.

If multiplying a negative number by a negative number, though, the answer will be positive. (-45) x (–4) would be 180, because you're taking away 45 sets of 4 somethings owed.

Division involving negative numbers is exactly the same. If you have 1 (or 3, or 5) negative numbers in your equation, the answer is going to be negative.

If there are an even number of negative numbers, the answer will be positive, because 2 negative numbers who come into contact will make a positive one, just like in multiplication.

Other than that, all operations done with negative numbers are exactly the same. As long as you're careful to keep the signs straight, you'll be all right.

# Division

Division is really a two-step process. The first step is guessing how many times one number fits into part of another. The second part is subtracting one number from another. If we are faced with the following equation, for example,

$$22\overline{)1234}$$

The first thing we need to do is to decide how many times 22 goes into 12. None, obviously. If that doesn't work, we need to add another digit and see how many times it goes into 123. Using rounding in conjunction with a little mental multiplication, you can see that the first digit of the answer is going to lie somewhere between 5 and 7. On the low end, 5 x 20 is 100, while 7 x 20 is 140, too high.

Using the EasyLearn Math principles of picking numbers apart and subtracting left to right, there's really no reason to write anything down on paper. Just keep tacking on numbers to the end of the answer and remember which number you need to bring down next.

Remember that as far as division or multiplication is concerned, you can multiply or divide both numbers involved by any number as long as you do the same thing to the other number. The answer will be the same as you would have gotten anyway. Remember that multiples of 10 are particularly easy to work with.

Fractions are really nothing more than unsolved division problems. When you see a fraction like 11/22, all it really means is that it represents the answer to 11 divided by 22, but that it was more convenient or time-effective to leave it in its present form. To get a real number out of a fraction, simply divide the top number by the bottom.

To convert "mixed" fractions, like 5-3/4, to decimal form, leave the whole number alone and divide the fraction through.

To add fractions, just make sure the number on the bottom is the same. If it is not, like 5/3 and 2/6, you will need to do some quick multiplication to make it so. Remember that 2/3 and 4/6 come out to be the same number, or percentage of a whole piece of something.

To add unruly fractions like 3/7 and 2/5, you need to seek the *lowest common denominator*. The denominator is actually what the number on the bottom is called, and the "lowest common" part of the phrase means that you want to find the smallest number that can be divided evenly by both 5 and 7. In many cases, you will need to multiply the two bottom numbers together to find the lowest common denominator.

Sometimes when you are working on a puzzle which has the same variable both on top of and on the bottom of the bar signifying "divide," (it is really the same bar as in fractions, just longer), you can cancel the two variables out, since you are both multiplying by and dividing by whatever that variable is. This is really just an extension of simplifying fractions. Here is an example:

$$\frac{4x}{12x}$$

We could cancel out the x's on both top and bottom, since it is really like saying 4/4, or 1.

To subtract fractions, you must also have a couple of fractions with the same denominator. Then just subtract the second top number from the first. 4/5 - 1/5 = 3/5.

Multiplying and dividing fractions is a different story. To multiply fractions, multiply the top number by the top number, and the bottom number by the bottom number. No common denominator is needed. 2/5 x 4/7 = 8/35.

To divide fractions, you need to cross-multiply, or multiply the top number of one by the bottom number of the other, and vice versa. 2/5 ÷ 4/3 = 6/20.

As we learned when "finishing" fractions to make decimals, 66/100 divides out to be .66. All that is required to make a percentage out of a decimal is to move the decimal point two spaces to the right and add the "%" sign.

You can add and subtract decimals without doing anything to them. That is, if 24 percent of customers think your company is the greatest thing on earth and 35 percent think it's merely pretty good, you can say that 59 percent of customers surveyed like the company.

To multiply and divide percentages, however, you must turn each of them back into decimals by moving the decimal point two places to the left.

When we say we are looking for the square root of 25, what we are saying is, find me a number which, when multiplied by itself, will equal 25. Obviously, the answer in this case is 5.

To find the square root of a number which is not a "square," try comparing the number to the nearest squares. If you need to figure out the square root of 70, for instance, remember that the square of 8 is 64 and the square of 9 is 81, so the square root of 70 lies somewhere in between. In this case, 70 is 6 away from 64 and 7 away from 81, so the true square root would be a little under halfway, maybe 8.4 or so.

If you have a pencil and paper handy, there is a somewhat more involved way of finding square roots to a greater degree of accuracy. Suppose you need to find the square root of a number like 85. You know that 9 x 9 is 81, too low, and 10 x 10 is 100, too high. Obviously the answer is going to be "9 something," so write down a 9. To get the rest of the answer, divide 85 by 9, which gives us 9.444444444 and so on, which we will treat as 9.44, just dropping the rest of the numbers because we round down (or truncate) for 4, and round up for 5 or more.

Next we add 9 and 9.44 and average them to get 9.22, which is our final guess at the answer. In fact, $(9.22)^2$ is 85.0084, so we are really pretty close.

To check your work for division, multiply the check digit of the answer by the check digit of the number doing the dividing. If you come up with the check digit of the number being divided, you're all set.

# Putting it all together

# Chapter 12

## Putting it all together

## Order is everything

Remember the order of operations. It is as follows, and it generally proceeds from left to right, top to bottom:

1) If you have the value of $x$, substitute it for $x$ everywhere in the problem.

2) Work inside parentheses first.

3) Resolve exponents and square roots—change $3^3$ to 27, for example.

4) Work multiplication and division within parentheses.

5) Work addition and subtraction within parentheses.

6) If parentheses have exponents on them, resolve those.

7) Multiply and divide parentheses above the division bar.

8) Add and subtract parentheses above the bar.

9) Repeat steps 1 through 8 below the bar, if required.

10) Divide the top by the bottom, if required.

There is a neat little mnemonic enhancer for remembering the order of operations, generally: "Please Excuse My Dear Aunt Sally." If you can remember this sentence, you can remember the order of operations. The initial letters in the words "Please Excuse My Dear Aunt Sally" stand for:

Parentheses

Exponents

Multiplication

Division

Addition

Subtraction

If you have an equation with an unknown quality, you can still solve it as long as you remember the Golden Rule of Equations:

***Doest Thou to One Side What Thou Doest to the Other***

That is, whether you multiply, divide, add or subtract from one side of the equation, make sure you do it to the other side too.

Functions are nothing but number factories, really. They have a manufacturing process all set up, and you just can just run any number through them.

The first part, $f(x) = x^3 + 2x^2 + 4x$, does nothing but tell us what processes in the factory have been set up. That is, when you plug in a number, the factory will cube it, take two times its square, and multiply it by four, then add the whole mess together.

The second part, f(4), just says, "In this particular case, we want to plug in the number 4."

# Story problems

The rules for solving story problems are as follows:

1) Read the story problem all the way through, then read it again. Decide what it is that the puzzle is asking you to do. What pieces of the puzzle do you have, and which ones are missing?

2) Physically cross out any unnecessary facts. In order to find the length of a fence, for example, you really do not need to know that the owner of the fence, Frank, has neighbors who are named Eric and Susie.

3) Draw a picture, if you can, and label the appropriate parts.

4) Set up the pieces to your puzzle in an appropriate way to provide the answer.

5) Perform the calculations, keeping in mind that your answer will not be simply 4, but "4 oranges," or whatever.

6) Look at the answer to see if it seems reasonable, under the conditions set down in the puzzle.

If 1 orange = 10 apples, can 15 oranges really equal 186 apples? Check your work with check digits, if appropriate.

Conversions are a common type of story problem. Before you are able to work a conversion problem, you need to know how many of one thing equal one of another thing. A story problem will generally tell you, and if you are working a problem from real life, you will know (or at least be able to look it up).

Here is an example from the book:

*Farmer Frank finds he is able to trade 17 pumpkins for one ounce of gold. If he grows 423 pumpkins this year, how much gold can he trade them for?*

The pieces of the puzzle we are provided with are that 17 pumpkins equal 1 ounce of gold, and that we have 423 pumpkins to get rid of. The only thing we are after here is how many ounces of gold equal 423 pumpkins?

One could, obviously, simply divide 423 by 17, which would leave 24.88 ounces of gold. But what we are really doing when we divide 423 by 17 is setting up the problem this way:

$$\frac{(423 \text{ pumpkins})}{1} \times \frac{1 \text{ oz. gold}}{17 \text{ pumpkins}}$$

When we multiply through, just like these were regular fractions (so top-times-top, bottom-times-bottom), we get:

$$\frac{423 \text{ pumpkins} \times \text{oz. gold}}{17 \text{ pumpkins}}$$

Since the "pumpkins" on top and the "pumpkins" on bottom are the same term, they cancel out (just like 4/4 = 1).

$$\frac{423 \text{ oz. gold}}{17}$$

# Everyday math

☆ The formula for figuring out a grocery bill in your head is:

(P1 + P2 + P3...) + [T x (P1 + P2 + P3...)]

Where P1, P2, and so on are the prices of the individual items, each rounded up to the nearest quarter (to reduce addition mistakes), and T stands for the tax rate where you live. The brackets, or [ ], function just like parentheses do, and as the order of operations tells us, you work from inside the parentheses to inside the brackets, to outside the brackets.

☆ Figuring out the bill in a restaurant when you have limited cash differs in two ways from adding up a grocery bill:

1) unless you wish to be thought of as a cheapskate, you must leave a tip of some sort, and

2) if you miscalculate you end up washing dishes for an hour.

Other than that, the calculations are pretty much the same. Here is the equation:

$$\{(P1 + P2 + P3...) + [T \times (P1 + P2 + P3...)]\} + G$$

Where P1 and so on are the prices of items you wish to order, T is tax, and G is gratuity, at whatever percentage you think the waiter deserves.

☆ To find out your hourly rate of pay (after taxes), simply take the amount of your paycheck and divide it by the number of hours you worked during the current pay period. Do this two or three weeks in a row (improving your sampling size to make sure the first check was not in error), and you should have a firm figure.

☆ Another time when "math on the fly" is useful is when you happen to see a sale at a favorite store. Say you had your eye on a CD player which normally retails for $300. This week, however, Larry's Loose Wires is having a blowout sale—everything in the store is 25% off. How much does the CD player cost now?

There are two ways to approach a problem like this:

1) you could multiply $300 by .75, as 100% - 25% = 75%,

2) you could multiply \$300 by .25, then subtract that answer from \$300. Either way you get the same answer: \$225.

☆ To figure out your gas mileage, simply record your car's odometer reading when you fill the tank. Then next time you fill the tank, take note of the new odometer reading, and subtract the old from the new. Then divide the total number of miles you put on since the last fill by the number of gallons (not the dollar amount) it took to fill your tank this time. The equation looks like this:

$$(M_{new} - M_{old}) \div G$$

Where M stands for miles on the odometer and G, in this case, stands for gallons.

☆ Here are the equations to figure out perimeters on common geometrical shapes:

Square: 4 x Length

Rectangle: 2 x (Length + Height)

Right Triangle: $L + H + [\text{sqr. root } (L^2 + H^2)]$

Circle: 2 x π x radius. (π is roughly equal to 3.14).

☆ Formulas for finding the area of common geometrical shapes:

To find the area of a square, measure one side and square the measurement.

To find the area of a rectangle, multiply two adjoining sides.

To find the area of a right triangle, multiply the two short sides and divide by 2.

To find the area of a circle, multiply the radius by itself, then by π.

# Math and your bank

☆ The formula for calculating simple interest is: $T = Prn$

T stands for total amount you now have.

P stands for the principal you started with.

r stands for the rate of interest you are receiving, in decimal form.

n stands for the number of years involved.

☆ Compound interest is calculated using the following equation:

$$T = P(1 + I)^n$$

The variables are defined in this manner:

T is the total amount of money we end up with, or the principal plus accumulated interest.

P is the principal.

I is the interest rate we get, divided by 100 because. we are translating from a percent to a decimal. For compound interest, we also divide the stated rate by the number of times per year the interest will be compounded—for quarterly compounding, we would divide by 4, for example.

n is the number of interest periods that we leave the money in the bank. If our deposit was compounded quarterly, for example, and we left the money in for a year, n would be 4.

☆ All installment loans and most credit cards should be treated as compound interest.

☆ Here is the formula for calculating mortgage payments:

$$P = (\text{Amount of Loan}) \times I \, (1 + I)^n$$

$$(1 + I)^n - 1$$

In this case, "P" stands for monthly payment, "I" for the interest rate in decimal form, and "n" for the number of payments in the loan.

☆ I recommend never purchasing the credit life and disability option on any loan, and that includes credit cards. There are three reasons for this:

One, the insurance is generally really overpriced.

Two, the purchase price of the insurance is stacked on top of your loan, so you are paying interest on that too!

Three, the insurance you purchase only covers the loan itself, not any other loans you have taken out. If you pay off all but $100 of your loan and then become disabled, the insurance pays only the remaining $100—no more.

Here is the formula for figuring out effective annual yield: $1000 (1 + I)k - 1000$, where you know our old friend "I," the decimal form of the quoted rate divided by the number of times per year interest is compounded, which "k" just happens to represent. This formula calculates the amount of interest generated at the effective rate if $1,000 is left in the account for a year.

☆ The formula for calculating the degree of appreciation on assets over the course of a year is:

$$\% \text{ Increase} = (P_1 - P_0) \div P_0 \times 100$$

$P_0$ is the original purchase price, $P_1$ is the current value.

Workbook

13

# Workbook

Work Area

## Addition:

| 1 | 2 | 3 | 4 |
|---|---|---|---|
| 6 | 4 | 8 | 4 |
| 3 | 9 | 9 | 4 |
| 1 | 1 | 8 | 6 |
| 4 | 7 | 2 | 3 |
| 2 | 6 | 3 | 1 |
| 7 | 6 | 2 | 7 |
| 8 | 2 | 1 | 6 |
| + 2 | + 3 | + 7 | + 2 |

| 5 | 6 | 7 |
|---|---|---|
| 8 | 6 | 3 |
| 6 | 2 | 6 |
| 6 | 3 | 2 |
| 3 | 4 | 8 |
| 2 | 7 | 4 |
| 9 | 9 | 1 |
| 1 | 2 | 7 |
| + 4 | + 1 | + 4 |

| 8 | 9 | 10 | 11 |
|---|---|---|---|
| 23 | 49 | 36 | 22 |
| + 54 | + 12 | + 97 | + 16 |

| 12 | 13 | 14 | 15 |
|---|---|---|---|
| 38 | 67 | 29 | 99 |
| + 44 | + 19 | + 72 | + 73 |

Work Area

16\. 468 + 193

17\. 112 + 362

18\. 256 + 362

19\. 147 + 211

20\. 983 + 429

21\. 602 + 429

22\. 229 + 182

23\. 3990 + 4267

24\. 7032 + 1977

25\. 6211 + 3728

26\. 6069 + 1124

27\. 1299 + 1327

28\. 6789 + 7271

29\. 41253 + 918

30\. 23747 + 39828

31\. 14728 + 43

32\. 89213 + 74800

33\. 32960 + 11203

34\. 91628 + 34807 + 12693 + 41877

35\. 21639 + 81222 + 40733 + 12793

36\. 91632 + 40103 + 22235 + 15401

37\. 86391 + 29201 + 43772 + 10245

38\. 12745 + 16938 + 40129 + 36155

39\. 45.37 + 98.01

40\. 36.98 + 12.73

41\. 37.12 + 4.98

Work Area

42. 5.62 + 18.12

43. 73.18 + 42.90

44. 20.00 + 12.34

45. 112.39 + 48.42 + 129.00 + 701.25

46. 169.53 + 142.07 + 329.98 + 802.41

47. 43.09 + 112.81 + 403.89 + 107.12

48. 223.12 + 401.18 + 4.06 + 719.30

49. 42.98 + 107.63 + 33.42 =

50. 103.29 + 42.12 + 4.09 =

51. 19.23 + 31.42 + 16.23 =

52. 10.77 + 22.12 + 403.55 =

53. 19.28 + 33.33 + 4.117 =

54. 73.221 + 10408 + 207.223 =

55. 69.08 + 72.721 + 22.75 =

56. 105.34 + 24.107 + 9.3333 =

57. 49.19 + 18.423 + 2.43 =

# Multiplication:

Work Area

| | | |
|---|---|---|
| 1. 29 x 7 | 2. 39 x 4 | 3. 22 x 7 |
| 4. 49 x 8 | 5. 53 x 4 | 6. 99 x 8 |
| 7. 87 x 4 | 8. 45 x 6 | 9. 18 x 5 |
| 10. 37 x 9 | 11. 82 x 8 | 12. 61 x 7 |
| 13. 19 x 3 | 14. 29 x 7 | 15. 49 x 23 |
| 16. 60 x 18 | 17. 33 x 12 | 18. 98 x 37 |
| 19. 72 x 27 | 20. 66 x 42 | 21. 82 x 14 |
| 22. 22 x 18 | 23. 38 x 47 | 24. 69 x 96 |
| 25. 84 x 47 | 26. 98 x 47 | 27. 37 x 12 |
| 28. 75 x 25 | 29. 86 x 44 | 30. 440 x 5 |
| 31. 612 x 5 | 32. 123 x 5 | 33. 912 x 5 |

Work Area

| | | |
|---|---|---|
| 34. 832 x 5 | 35. 306 x 5 | 36. 777 x 5 |
| 37. 223 x 15 | 38. 891 x 15 | 39. 107 x 15 |
| 40. 922 x 15 | 41. 432 x 15 | 42. 622 x 15 |
| 43. 707 x 15 | 44. 489 x 25 | 45. 661 x 25 |
| 46. 473 x 25 | 47. 512 x 25 | 48. 909 x 25 |
| 49. 823 x 25 | 50. 117 x 25 | 51. 47 x 50 |
| 52. 96 x 50 | 53. 38 x 50 | 54. 129 x 50 |
| 55. 432 x 50 | 56. 912 x 50 | 57. 1063 x 50 |
| 58. 53 x .17 | 59. 10.24 x .24 | 60. 9.62 x 13 |
| 61. 18.43 x .08 | 62. 19.69 x .04 | 63. 23.16 x .75 |

Work Area

64 $4^3$ 65 $10^8$ 66 $9^5$

67 $14^2$ 68 $7^1$ 69 $3^3$

70 $8^4$ 71 $9^3$

72 $2.05 \times 10^3$ 73 $1.179 \times 10^5$

74 $3.201 \times 10^{-4}$ 75 $1.116 \times 10^{-8}$

76 $3.14 \times 10^1$ 77 $4.20 \times 10^0$

78 $117 \times 10^{-2}$ 79 $401 \times 10^{-3}$

## *Factor:*

80 14 81 15 82 27

83 84 84 87 85 90

86 162 87 203 88 445

89 916 90 222 91 360

Work Area

# Subtraction:

1. $\begin{array}{r} 413 \\ -\ 107 \\ \hline \end{array}$
2. $\begin{array}{r} 227 \\ -\ 113 \\ \hline \end{array}$
3. $\begin{array}{r} 451 \\ -\ 22 \\ \hline \end{array}$
4. $\begin{array}{r} 508 \\ -\ 38 \\ \hline \end{array}$
5. $\begin{array}{r} 612 \\ -\ 147 \\ \hline \end{array}$
6. $\begin{array}{r} 901 \\ -\ 399 \\ \hline \end{array}$
7. $\begin{array}{r} 612 \\ -\ 39 \\ \hline \end{array}$
8. $\begin{array}{r} 451 \\ -\ 112 \\ \hline \end{array}$
9. $\begin{array}{r} 637 \\ -\ 201 \\ \hline \end{array}$
10. $\begin{array}{r} 891 \\ -\ 14 \\ \hline \end{array}$
11. $\begin{array}{r} 917 \\ -\ 38 \\ \hline \end{array}$
12. $\begin{array}{r} 723 \\ -\ 27 \\ \hline \end{array}$
13. $\begin{array}{r} 119 \\ -\ 201 \\ \hline \end{array}$
14. $\begin{array}{r} 338 \\ -\ 600 \\ \hline \end{array}$
15. $\begin{array}{r} 391 \\ -\ 712 \\ \hline \end{array}$
16. $\begin{array}{r} 493 \\ -\ 508 \\ \hline \end{array}$
17. $\begin{array}{r} 327 \\ -\ 618 \\ \hline \end{array}$
18. $\begin{array}{r} 112 \\ -\ 300 \\ \hline \end{array}$
19. 429 – 113 =
20. 398 – 107 =
21. 612 – 457 =
22. 107 – 43 =
23. 711 – 17 =
24. 1,031 – 227 =
25. 1,236 – 908 =
26. 723 – 92 =

## *Solve:*

Work Area

27. 1630 – (–29) =
28. 48 – (–50) =
29. 56 – (–93) =
30. 18 – (–32) =
31. 64 + (–12) =
32. 169 + (–12) =
33. 383 + (–72) =
34. 99 + (–18) =
35. 36 (–12) =
36. –18 (13) =
37. –12 (–16) =
38. –7 (–36) =
39. $\begin{array}{r} 98.01 \\ -\ 45.37 \\ \hline \end{array}$
40. $\begin{array}{r} 36.98 \\ -\ 12.73 \\ \hline \end{array}$
41. $\begin{array}{r} 37.96 \\ -\ 4.12 \\ \hline \end{array}$
42. $\begin{array}{r} 12.18 \\ -\ 16.42 \\ \hline \end{array}$
43. $\begin{array}{r} 112.39 \\ -\ 48.42 \\ \hline \end{array}$
44. $\begin{array}{r} 169.53 \\ -\ 32.47 \\ \hline \end{array}$
45. $\begin{array}{r} 403.89 \\ -\ 107.12 \\ \hline \end{array}$
46. $\begin{array}{r} 223.12 \\ -\ 162.41 \\ \hline \end{array}$

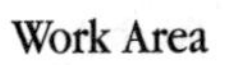


# Division

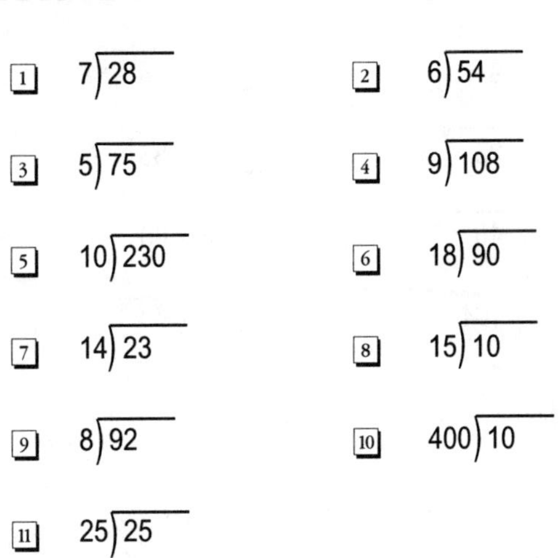


## *Change to Decimals:*

12 $\frac{1}{4}$

13 $\frac{3}{8}$

14 $\frac{6}{12}$

15 $\frac{4}{9}$

16 $\frac{3}{2}$

17 $\frac{8}{12}$

18 $\frac{7}{8}$

19 $\frac{1}{3}$

20 $\frac{4}{3}$

21 $\frac{12}{8}$

22 $\frac{9}{6}$

23 $\frac{7}{5}$

24 $5\frac{1}{2}$

25 $6\frac{1}{3}$

26 $2\frac{3}{4}$

27 $3\frac{5}{7}$

## *Estimate:*

Work Area

28 $\sqrt{9}$

29 $\sqrt{16}$

30 $\sqrt{25}$

31 $\sqrt{36}$

32 $\sqrt{46}$

33 $\sqrt{83}$

34 $\sqrt{104}$

35 $\sqrt{63}$

## *Add:*

36 $\frac{4}{5} + \frac{7}{5} =$

37 $\frac{5}{4} + \frac{3}{2} =$

38 $\frac{3}{4} + \frac{1}{3} =$

39 $\frac{12}{7} + \frac{2}{5} =$

40 $\frac{9}{10} + \frac{7}{8} =$

41 $\frac{11}{8} + \frac{8}{11} =$

42 $10\frac{1}{3} + \frac{6}{5} =$

43 $\frac{3}{4} + \frac{1}{8} =$

44 $7\frac{5}{6} + 3\frac{1}{2} =$

45 $4\frac{7}{6} + 5\frac{1}{6} =$

## *Multiply:*

46 $\frac{6}{8} \cdot \frac{3}{4} =$

47 $\frac{4}{6} \cdot \frac{1}{3} =$

48 $\frac{5}{6} \cdot \frac{2}{3} =$

49 $\frac{1}{2} \cdot \frac{5}{8} =$

50 $\frac{1}{4} \cdot \frac{1}{8} =$

51 $\frac{1}{6} \cdot \frac{2}{3} =$

52 $\frac{7}{8} \cdot \frac{11}{12} =$

53 $\frac{6}{7} \cdot \frac{3}{4} =$

54 $\frac{10}{7} \cdot \frac{4}{5} =$

55 $\frac{11}{7} \cdot \frac{5}{6} =$

Work Area

## *Subtract:*

56 $\frac{6}{8} - \frac{3}{4} =$

57 $\frac{6}{7} - \frac{3}{4} =$

58 $\frac{7}{8} - \frac{11}{12} =$

59 $\frac{9}{4} - \frac{1}{3} =$

60 $\frac{10}{7} - \frac{4}{5} =$

61 $\frac{7}{8} - \frac{2}{6} =$

62 $4\frac{1}{5} - 3\frac{1}{3} =$

63 $5\frac{1}{4} - \frac{3}{5} =$

64 $3\frac{7}{8} - 2\frac{5}{9} =$

65 $3\frac{3}{7} - 2\frac{2}{3} =$

## *Divide:*

66 $\frac{4}{3} \div \frac{2}{3} =$

67 $\frac{1}{3} \div \frac{1}{2} =$

68 $\frac{4}{8} \div \frac{1}{5} =$

69 $\frac{5}{7} \div \frac{6}{2} =$

70 $\frac{9}{11} \div \frac{1}{7} =$

71 $\frac{3}{8} \div \frac{5}{6} =$

72 $\frac{7}{9} \div \frac{1}{3} =$

73 $\frac{6}{7} \div \frac{1}{4} =$

74 $-312 \div (-4) =$

75 $-69 \div (-3) =$

76 $-49 \div (-7) =$

77 $306 \div (-12) =$

78 $-465 \div (15) =$

79 $510 \div (-10) =$

80 $-35 \div (-5) =$

81 $175 \div (25) =$

82 $450 \div (-75) =$

83 $759 \div (-33) =$

84 $-96 \div (6) =$

85 $156 \div (-4) =$

# Putting It All Together

Work Area

## *Solve:*

1. $(3)(3) - (4)(6) =$

2. $(4)(2) - (6)(8) =$

3. $(37 + 4)(12 + 19) =$

4. $(5)(3 - 1)(2 + 1) =$

5. $12^2 + 3^3 - 4^3 + 2^2 =$

6. $9^2 - 3^2 \div 3 + 5^2 =$

7. $(36 \div 6)^2 + (48)(2) =$

8. $(25 \div 5)^2 + (30 \div 2) =$

9. $\frac{(13 + 11)(4)}{2}$

10. $\frac{(3)(10^2)}{30}$

11. $\frac{(28 \div 7)^2 + 13}{\sqrt{36}}$

12. $\frac{(21 - 17)^3 + (-16)}{\sqrt{64}}$

## *x = 4; Solve:*

13. $x^2 + x + 1/x =$

14. $2x + x^3 + 2/x =$

15. $37x + x + 4 =$

16. $15x + 2x^2 - 10 =$

17. $x^3 + 3x^2 - x =$

18. $x^3 - 2x^2 + x =$

Work Area

19 $\frac{(x^2 + 2)(x^3 - 12)}{3}$

20 $\frac{(x^2 - 2)(x^2 + 2)}{12}$

21 $\frac{42x + 3x - (14 + x)}{4}$

22 $\frac{(5x^2 - 3x) \div (2x + 2)}{x}$

## *x = 5, y = 7; Solve:*

23 $(x + y)(x - y) =$

24 $2x + yx - 3y =$

25 $\frac{(x^2 + y) - (y^3)}{y}$

26 $\frac{(x^2 + y^2)(x^2 - y^2)}{x^2}$

27 $\frac{3x^3 + 2y}{3y}$

28 $\frac{2x^2 - 3xy}{2x}$

## *x = 13, y = 3/7 ; Solve:*

29 $x^2 + y^2 =$

30 $x^2 - 7y =$

31 $3x^2 + 2y^2 =$

32 $x^2/6 + 3y^2 =$

33 $(xy)(^1/_2 y) =$

34 $(x - 4) \div 5y^2 =$

## *Solve for x:*

35 $3x^2 + 13 = 0$

36 $2x^2 - 4x = 0$

37 $(4x)\ 3 + 4 = 0$

38 $4x = 12$

39 $10x \div 5 = 0$

40 $3x^2 - 6 = 6$

41 $\frac{5x^2 + 3x}{X} = 0$

42 $\frac{2x^2 - x}{4} = 0$

# Story Problems

Work Area

## *Draw pictures, if applicable, and solve:*

1) Hugo White is taking a vacation to Gonesh, where the local currency, the Thud, is equal to 27 U.S. Dollars. Hugo is going to take his life savings of $432; how many Thuds can he expect to receive in exchange?

2) When Hugo arrives in Gonesh, his cab driver offers him 3 Thuds for $75, rather than the official exchange rate of 1 Thud to 27 Dollars. Assuming Hugo wants the most for his money, should he take the cabbie's offer? How many extra Thuds, if any, will he receive?

3) Farmer Frank needs to take a trip to town; his 1936 Ford pickup has a top speed of 30 miles per hour. Assuming town is 10 miles away, how long would it take Frank to get to town in an emergency? Assuming Frank's wildcatting days are done, and he drives at the saner speed of 20 mph, how long will it take him to get to town?

Work Area

4) When Frank gets to town, he buys 30 50-lb. bags of fertilizer. In order to get a discount, he loads the bags into his pick-up himself. When he finishes, how many pounds has he lifted? Express this number in kilograms (1 kg = 2.2 lbs).

5) Susie, Frank's neighbor, runs a barter club. She and the other members of the club have agreed on the following exchange rates:

30 cucumbers = 1 bushel of sweet corn
12 lbs. of honey = 1 bushel of sweet corn
10 bushels of sweet corn = 1 side of beef
2 sides of beef = 1 oz. of gold
2 oz. of gold = 1 horse

Farmer Frank had a bumper crop of cucumbers this year, and he needs to get rid of them before they spoil. He has 300 cucumbers to get rid of.

How much corn could he get in exchange?

How much honey could he get?

Can he afford a side of beef?

If Frank has a horse he'd like to get rid of, how many sides of beef can he get?

If, for some strange reason, Frank wanted to buy more cucumbers, how many could he get for the horse?

Work Area

6) Hilda and Jim are looking for a good deal on a certificate of deposit. Their bank is offering 6 month CD's at 3.9% and 12 month CD's at 4.4%. Assuming they have $3500 to invest, how much interest will they earn on the 6 month CD? The 12 month?

7) The bank manager tells Hilda and Jim that if they wait a week, the bank will be offering a 7 month CD at 6%. If they take this deal, can they earn more interest in 7 months than they could in 12 months at 4.4%?

8) If light travels 186,000 miles per second (which it does), how far does it travel in a week?

9) The volume of water in the earth's oceans is approximately 1 x 109 cubic kilometers. Given the following conversions, how many Pepsi cans could the oceans fill?

- 1 Pepsi can holds 355 milliliters (ml)
- 1 milliliter = 1 cubic centimeter ($cm^3$)
- 1 cubic kilometer ($km^3$) = 1,000,000 $cm^3$

Work Area

# Math In Real Life

1) On your way home from work, you need to pick up a few things. You only have a $20 bill, so you need to keep track of prices as you go. The sales tax rate in your state is 6% and it applies to all items purchased. You want to buy coffee ($4.59), eggs ($1.09), milk ($3.09), bread ($1.19), a newspaper ($.35), and a bag of dog food ($8.79). Can you buy all of these items? How much money will you have left over, or be short by?

2) A store sells three different jars of spaghetti sauce. One is 16 ounces for $1.79, one is 28 ounces for $2.29, and one is 64 ounces for $6.59. Which is the best deal? Express the cost/ounce ratio for each.

3) You go out for dinner with some friends, and you agree to bet that whoever guesses closest to the total bill eats for free. Here is a list of what everybody eats:

| | | |
|---|---|---|
| Bill: | | |
| | Ribeye steak | $12.95 |
| | Coffee | $1.25 |
| | Apple Pie | $2.25 |
| Mary: | | |
| | Chicken Dinner | $6.95 |
| | Soda | $.85 |
| Sue: | | |
| | New York Strip | $13.95 |
| | Milk | $1.15 |
| | Strawberry Pie | $2.25 |
| You: | | |
| | Chef Salad | $4.95 |
| | Chicken Wings | $3.95 |
| | Coffee | $1.25 |

Work Area

Assuming a tax rate of 6%, estimate the total cost of dinner (in your head). Then, add up the actual prices, including tax, and see how close you came.

4) Bruce leaves home and drives at 35 mph for 10 minutes. Then he gets on the freeway and drives 80 mph for 20 minutes. When he gets pulled over by a state trooper, the car is parked for 10 minutes while he gets a ticket. Then he resumes driving at 55 mph. At the end of an hour, how far has Bruce traveled?

5) Farmer Frank is building a corral. He has already sunk posts for a fence; now, he needs to connect them with boards. He wants three rows of boards all the way around the corral. The corral is square and each side is 70 feet. How many feet of boards will he need?

6) Assuming the boards cost $.35 cents per foot, how much will the project cost him?

7) Now Frank decides he wants to run one strand of electric fence wire 6 inches inside the boards. How much wire will he need? (Hint - The perimeter will be shorter by 12 inches per side.)

Work Area

8) If wire is $.03 per foot, what is the total cost of the fence?

9) Jerry and Sheila have saved $3,000. Their bank offers a 1-year CD at 5.5%. At thc end of a year, how much interest will they have earned?

10) If they choose a 5-year CD at 6.8%, how much will they have earned when the certificate has matured?

11) What is the area of a circle that measures 6 inches across? What is the perimeter?

12) It is roughly 1800 miles from Minneapolis to San Francisco. State this distance in light years. (Hint: A light year is the distance light travels in a year. Light travels 1.86 x 105 miles per second.)

# Math And Your Bank

Work Area

1) Ron is trying to decide where to invest $5,000 he has saved. One bank offers a 1-year CD, compounded monthly, at 5.6%. Another offers a 1-year CD, compounded quarterly, at 5.7%. Calculate the interest generated on each. Which is the better deal?

2) How much money would be saved over the course of a 30 year mortgage if $150,000 was borrowed at 9 percent, rather than 10 percent?

3) What is the effective annual yield of 18% interest which is compounded monthly?

4) Dalia bought some pharmaceutical stock a year ago, when it cost $18.25 per share. It is now worth $20.00 per share. Is this stock performing better than a quarterly-compounding CD at 4.5 percent?

Work Area

5) If you found your dream house for $300,000 and managed to pay 10 percent as a downpayment, what would your monthly payment be if the rate was 11 percent and the term was 30 years? What would it be at a 20 year term? A 10 year term?

6) Frank is buying a new Hupmobile for $25,000. How much cheaper would it be for him to borrow the money on a simple interest, single payment note at 9%, rather than an installment loan? Assume both loans have a term of four years.

7) Mary has a credit card which charges 17.5%, compounded monthly. If she currently carries an average balance of $500, how much can she save each year by paying off her card in full each month?

# Answers

14

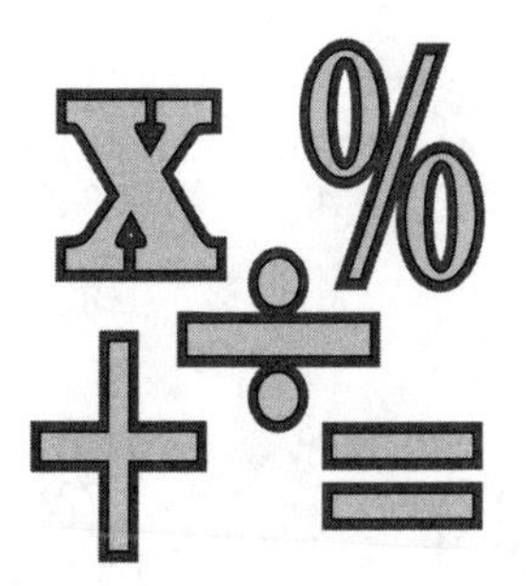

# Chapter 14

## Answers

### Addition

1.
```
   6
   3
   1
   4
   2
   7
   8
 + 2
  33
```

2.
```
   4
   9
   1
   7
   6
   6
   2
 + 3
  38
```

3.
```
   8
   9
   8
   2
   3
   2
   1
 + 7
  40
```

4.
```
   4
   4
   6
   3
   1
   7
   6
 + 2
  33
```

5.
```
   8
   6
   6
   3
   2
   9
   1
 + 4
  39
```

6.
```
   6
   2
   3
   4
   7
   9
   2
 + 1
  34
```

7.
```
   3
   6
   2
   8
   4
   1
   7
 + 4
  35
```

8.
```
   23    5
  +54    0
   77    5
```

9.
```
   49    4
 + 12    3
   61    7
```

10.
```
   36    0      36
 + 97    7 =  +100
  133    7     136
                -3
               133
```

**11**

| | | |
|---|---|---|
| 22 | 4 | |
| + 16 | 7 | |
| 38 | 2 | |

**12**

| | | |
|---|---|---|
| 38 | 2 | 40 |
| + 44 | 8 = | + 44 |
| 82 | 1 | 84 |
| | | – 2 |
| | | 82 |

**13**

| | | |
|---|---|---|
| 67 | 4 | 67 |
| + 19 | 1 = | + 20 |
| 86 | 5 | 87 |
| | | – 1 |
| | | 86 |

**14**

| | | |
|---|---|---|
| 29 | 2 | 30 |
| + 72 | 0 = | + 72 |
| 101 | 2 | 102 |
| | | – 1 |
| | | 101 |

**15**

| | | |
|---|---|---|
| 99 | 0 | 100 |
| + 73 | 1 = | + 73 |
| 172 | 1 | 173 |
| | | – 1 |
| | | 172 |

**16**

| | | |
|---|---|---|
| 468 | 0 | 468 |
| + 193 | 4 = | + 200 |
| 661 | 4 | 668 |
| | | – 7 |
| | | 661 |

**17**

| | | |
|---|---|---|
| 112 | 4 | 100 |
| + 362 | 2 = | + 362 |
| 474 | 6 | 462 |
| | | + 12 |
| | | 474 |

**18**

| | | |
|---|---|---|
| 256 | 4 | 250 |
| + 362 | 2 = | + 360 |
| 618 | 6 | 610 |
| | | + 8 |
| | | 618 |

**19**

| | | |
|---|---|---|
| 147 | 3 | |
| + 211 | 4 | |
| 358 | 7 | |

**20**

| | | |
|---|---|---|
| 983 | 2 | 1000 |
| + 429 | 6 = | + 429 |
| 1412 | 8 | 1429 |
| | | – 17 |
| | | 1412 |

**21**

| | | |
|---|---|---|
| 602 | 8 | 600 |
| + 429 | 6 = | + 429 |
| 1031 | 5 | 1029 |
| | | + 2 |
| | | 1031 |

**22**

| | | |
|---|---|---|
| 229 | 4 | 229 |
| + 182 | 2 = | + 200 |
| 411 | 6 | 429 |
| | | – 18 |
| | | 411 |

**23**

| | | |
|---|---|---|
| 3990 | 3 | |
| + 4267 | 1 | |
| 8257 | 4 | |

**24**

| | | |
|---|---|---|
| 7032 | 3 | 7000 |
| + 1977 | 6 = | + 1977 |
| 9009 | 0 | 8977 |
| | | + 32 |
| | | 9009 |

**25**

| | | |
|---|---|---|
| 6211 | 1 | |
| + 3728 | 2 | |
| 9939 | 3 | |

**26**

| | |
|---|---|
| 6069 | 3 |
| + 1124 | 8 |
| 7193 | 2 |

**27**

| | | |
|---|---|---|
| 1299 | 3 | 1300 |
| + 1327 | 4 = | + 1327 |
| 2626 | 7 | 2627 |
| | | − 1 |
| | | 2626 |

**28**

| | |
|---|---|
| 6789 | 3 |
| + 7271 | 8 |
| 14060 | 2 |

**29**

| | | |
|---|---|---|
| 41523 | 6 | 41523 |
| + 918 | 0 = | + 900 |
| 42441 | 6 | 42423 |
| | | + 18 |
| | | 42441 |

**30**

| | |
|---|---|
| 23747 | 5 |
| + 39828 | 3 |
| 63575 | 8 |

**31**

| | | |
|---|---|---|
| 14728 | 4 | 14728 |
| + 43 | 7 = | + 50 |
| 14771 | 2 | 14778 |
| | | − 7 |
| | | 14771 |

**32**

| | |
|---|---|
| 89213 | 5 |
| + 74800 | 1 |
| 164013 | 6 |

**33**

| | |
|---|---|
| 32960 | 2 |
| + 11203 | 7 |
| 44163 | 0 |

**34**

| | |
|---|---|
| 91628 | 8 |
| 34807 | 4 |
| 12693 | 3 |
| + 41877 | 0 |
| 181005 | 6 |

**35**

| | |
|---|---|
| 21639 | 3 |
| 81222 | 6 |
| 40733 | 8 |
| + 12793 | 4 |
| 156387 | 3 |

**36**

| | |
|---|---|
| 91632 | 3 |
| 40103 | 8 |
| 22235 | 5 |
| + 15401 | 2 |
| 169371 | 0 |

**37**

| | |
|---|---|
| 86391 | 0 |
| 29201 | 5 |
| 43772 | 5 |
| + 10245 | 3 |
| 169609 | 4 |

**38**

| | |
|---|---|
| 12745 | 1 |
| 16938 | 0 |
| 40129 | 7 |
| + 36155 | 2 |
| 105967 | 1 |

39

| | | |
|---|---|---|
| 45.37 | 1 | |
| + 98.01 | 0 | |
| 143.38 | 1 | |

40

| | | |
|---|---|---|
| 36.98 | 8 | 37.00 |
| + 12.73 | 4 = | + 12.73 |
| 49.71 | 3 | 49.73 |
| | | − .02 |
| | | 49.71 |

41

| | | |
|---|---|---|
| 37.12 | 4 | 37.12 |
| + 4.98 | 3 = | + 5.00 |
| 42.10 | 7 | 42.12 |
| | | − .02 |
| | | 42.10 |

42

| | | |
|---|---|---|
| 5.62 | 4 | 5.62 |
| + 18.12 | 3 = | + 18.00 |
| 23.74 | 7 | 23.62 |
| | | + .12 |
| | | 23.74 |

43

| | |
|---|---|
| 73.18 | 1 |
| + 42.90 | 6 |
| 116.08 | 7 |

44

| | |
|---|---|
| 20.00 | 2 |
| + 12.34 | 1 |
| 32.34 | 3 |

45

| | |
|---|---|
| 112.39 | 7 |
| 48.42 | 0 |
| 129.00 | 3 |
| + 701.25 | 6 |
| 991.06 | 7 |

46

| | |
|---|---|
| 169.53 | 6 |
| 142.07 | 5 |
| 329.98 | 4 |
| + 802.41 | 6 |
| 1443.99 | 3 |

47

| | |
|---|---|
| 43.09 | 7 |
| 112.81 | 4 |
| 403.89 | 6 |
| + 107.12 | 2 |
| 666.91 | 1 |

48

| | |
|---|---|
| 223.12 | 1 |
| 401.18 | 5 |
| 4.06 | 1 |
| + 719.30 | 2 |
| 1347.66 | 0 |

49

| | |
|---|---|
| 42.98 | 5 |
| 107.63 | 8 |
| + 33.42 | 3 |
| 184.03 | 7 |

50

| | |
|---|---|
| 103.29 | 6 |
| 42.12 | 0 |
| + 4.09 | 4 |
| 149.50 | 1 |

51

| | |
|---|---|
| 19.23 | 6 |
| 31.42 | 1 |
| + 16.23 | 3 |
| 66.88 | 1 |

52

| | |
|---|---|
| 10.77 | 6 |
| 22.12 | 7 |
| + 403.55 | 8 |
| 436.44 | 3 |

53

| | |
|---|---|
| 19.28 | 2 |
| 33.33 | 3 |
| + 4.117 | 4 |
| 56.727 | 0 |

54

| | |
|---|---|
| 73.221 | 6 |
| 10408 | 4 |
| + 207.223 | 7 |
| 10688.444 | 8 |

55

| | |
|---|---|
| 69.08 | 5 |
| 72.721 | 1 |
| + 22.75 | 7 |
| 164.551 | 4 |

56

| | |
|---|---|
| 105.34 | 4 |
| 24.107 | 5 |
| + 9.3333 | 3 |
| 138.7803 | 3 |

57

| | |
|---|---|
| 49.19 | 5 |
| 18.423 | 0 |
| + 2.43 | 0 |
| 70.043 | 5 |

# Multiplication:

1

| | | |
|---|---|---|
| 29 | 2 | (20 x 7) + ( 9 x 7) |
| x 7 | 7 = | 140 + 63 |
| 203 | 5 | 203 |

2

| | | |
|---|---|---|
| 39 | 3 | (30 x 4) + (9 x 4) |
| x 4 | 4 = | 120 + 36 |
| 156 | 3 | 156 |

3

| | | |
|---|---|---|
| 22 | 4 | (20 x 7) + (2 x 7) |
| x 7 | 7 = | 140 + 14 |
| 154 | 1 | 154 |

4

| | | |
|---|---|---|
| 49 | 4 | (40 x 8) + (9 x 8) |
| x 8 | 8 = | 320 + 72 |
| 392 | 5 | 392 |

5

| | | |
|---|---|---|
| 53 | 8 | (50 x 4) + (3 x 4) |
| x 4 | 4 = | 200 + 12 |
| 212 | 5 | 212 |

6

| | | |
|---|---|---|
| 99 | 0 | (90 x 8) + (9 x 8) |
| x 8 | 8 = | 720 + 72 |
| 792 | 0 | 792 |

| | | | | |
|---|---|---|---|---|
| 7 | 87 | 6 | | (80 x 4) + (7 x 4) |
| | x 4 | 4 | = | 320 + 28 |
| | 348 | 6 | | 348 |
| 8 | 45 | 0 | | (40 x 6) + (5 x 6) |
| | x 6 | 6 | = | 240 + 30 |
| | 270 | 0 | | 270 |
| 9 | 18 | 0 | | (10 x 5) + (8 x 5) |
| | x 5 | 5 | = | 50 + 40 |
| | 90 | 0 | | 90 |
| 10 | 37 | 0 | | (30 x 9) + (7 x 9) |
| | x 9 | 0 | = | 270 + 63 |
| | 333 | 0 | | 333 |
| 11 | 82 | 1 | | (80 x 8) + (2 x 8) |
| | x 8 | 8 | = | 640 + 16 |
| | 656 | 8 | | 656 |
| 12 | 61 | 7 | | (60 x 7) + (1 x 7) |
| | x 7 | 7 | = | 420 + 7 |
| | 427 | 4 | | 427 |
| 13 | 19 | 1 | | (10 x 3) + (9 x 3) |
| | x 3 | 3 | = | 30 + 27 |
| | 57 | 3 | | 57 |
| 14 | 29 | 2 | | (20 x 7) + (9 x 7) |
| | x 7 | 7 | = | 140 + 63 |
| | 203 | 5 | | 203 |
| 15 | 49 | 4 | | 50 |
| | x 23 | 5 | = | x 23 |
| | 1127 | 2 | | 1150 |
| | | | | – 23 |
| | | | | 1127 |

| 16 | | | |
|---|---|---|---|
| 60 | 6 | | (60 x 10) + (60 x 8) |
| x 18 | 0 | = | 600 + 480 |
| 1080 | 0 | | 1080 |

| 17 | | | |
|---|---|---|---|
| 33 | 6 | | (33 x 10) + (33 x 2) |
| x 12 | 3 | = | 330 + 66 |
| 396 | 0 | | 396 |

| 18 | | | |
|---|---|---|---|
| 98 | 8 | | 100 |
| x 37 | 1 | = | x 37 |
| 3626 | 8 | | 3700 |
| | | | – 74 |
| | | | 3626 |

| 19 | | | |
|---|---|---|---|
| 72 | 0 | | 70 |
| x 27 | 0 | = | x 27 |
| 1944 | 0 | | 1890 |
| | | | + 54 |
| | | | 1944 |

| 20 | | | |
|---|---|---|---|
| 66 | 3 | | (66 x 40) + (66 x 2) |
| x 42 | 6 | = | 2640 + 132 |
| 2772 | 0 | | 2772 |

| 21 | | | |
|---|---|---|---|
| 82 | 1 | | (80 x 14) + (2 x 14) |
| x 14 | 5 | = | 1120 + 28 |
| 1148 | 5 | | 1148 |

| 22 | | | |
|---|---|---|---|
| 22 | 4 | | (20 x 18) + (2 x 18) |
| x 18 | 0 | = | 360 + 36 |
| 396 | 0 | | 396 |

| 23 | | | |
|---|---|---|---|
| 38 | 2 | | 40 |
| x 47 | 2 | = | x 47 |
| 1786 | 4 | | 1880 |
| | | | – 94 |
| | | | 1786 |

| 24 | | | |
|---|---|---|---|
| 69 | 6 | | 70 |
| x 96 | 6 | = | x 96 |
| 6624 | 0 | | 6720 |
| | | | – 96 |
| | | | 6624 |

| 25 | | | |
|---|---|---|---|
| 84 | 3 | | 84 |
| x 47 | 2 | = | x 50 |
| 3948 | 6 | | 4200 |
| | | | – 252 |
| | | | 3948 |

| 26 | | | |
|---|---|---|---|
| 98 | 8 | | 100 |
| x 47 | 2 | = | x 47 |
| 4606 | 7 | | 4700 |
| | | | – 94 |
| | | | 4606 |

| | | | |
|---|---|---|---|
| 27 | 37<br>x 12<br>444 | 1<br>3 =<br>3 | (37 x 10) + ( 37 x 2)<br>370 + 74<br>444 |
| 28 | 75<br>x 25<br>1875 | 3<br>7 =<br>3 | (75 x 20) + (75 x 5)<br>1500 + 375<br>1875 |
| 29 | 86<br>x 44<br>3784 | 5<br>8 =<br>4 | (86 x 40) + ( 86 x 4)<br>3440 + 344<br>3784 |
| 30 | 440<br>x 5<br>2200 | 8<br>5 =<br>4 | (400 x 5) + (40 x 5)<br>2000 + 200<br>2200 |
| 31 | 612<br>x 5<br>3060 | 0<br>5 =<br>0 | (600 x 5) + (12 x 5)<br>3000 + 60<br>3060 |
| 32 | 123<br>x 5<br>615 | 6<br>5 =<br>3 | (120 x 5) + (3 x 5)<br>600 + 15<br>615 |
| 33 | 912<br>x 5<br>4560 | 3<br>5 =<br>6 | (900 x 5) + ( 12 x 5)<br>4500 + 60<br>4560 |
| 34 | 832<br>x 5<br>4160 | 4<br>5 =<br>2 | (800 x 5) + (32 x 5)<br>4000 + 160<br>4160 |
| 35 | 306<br>x 5<br>1530 | 0<br>5 =<br>0 | (300 x 5) + (6 x 5)<br>1500 + 30<br>1530 |
| 36 | 777<br>x 5<br>3885 | 3<br>5 =<br>6 | (700 x 5) + (70 x 5) + (7 x 5)<br>3500 + 350 + 35<br>3885 |

37
```
 223    7      223
x 15    6 =   x 10
3345    6     2230
            + 1115
              3345
```

38
```
   891    0      891
  x 15    6 =   x 10
13,365    0     8910
              + 4455
              13,365
```

39
```
 107    8      107
x 15    6 =   x 10
1605    3     1070
             + 535
              1605
```

40
```
   922    4      922
  x 15    6 =   x 10
13,830    6     9220
              + 4610
              13,830
```

41
```
 432    0      432
x 15    6 =   x 10
6480    0     4320
            + 2160
              6480
```

42
```
 622    1      622
x 15    6 =   x 10
9330    6     6220
            + 3110
              9330
```

43
```
   707    5      707
  x 15    6 =   x 10
10,605    3     7070
              + 3535
              10,605
```

44
```
   489    3      489
  x 25    7 =   x 10
12,225    3     4890
              + 2445
                7335
              + 4890
              12,225
```

45
```
   661    4      661
  x 25    7 =   x 10
16,525    1     6610
              + 3305
                9915
              + 6610
              16,525
```

46
```
   473    5      473
  x 25    7 =   x 10
11,825    8     4730
              + 2365
                7095
              + 4730
              11,825
```

47
```
   512    8      512
  x 25    7 =   x 10
12,800    2     5120
              + 2560
                7680
              + 5120
              12,800
```

48
```
   909    0      909
  x 25    7 =   x 10
22,725    0     9090
              + 4545
              13,635
              + 9090
              22,725
```

| | | | |
|---|---|---|---|
| 49 | 823 | 4 | 823 |
| | x 25 | 7 = | x 10 |
| | 20,575 | 1 | 8230 |
| | | | + 4115 |
| | | | 12345 |
| | | | + 8230 |
| | | | 20,575 |

| | | | |
|---|---|---|---|
| 50 | 117 | 0 | 117 |
| | x 25 | 7 = | x 10 |
| | 2925 | 0 | 1170 |
| | | | + 585 |
| | | | 1755 |
| | | | + 1170 |
| | | | 2925 |

| | | | |
|---|---|---|---|
| 51 | 47 | 2 | 47 |
| | x 50 | 5 = | x 100 |
| | 2350 | 1 | 2 / 4700 = 2350 |

| | | | |
|---|---|---|---|
| 52 | 96 | 6 | 96 |
| | x 50 | 5 = | x 100 |
| | 4800 | 3 | 2 / 9600 = 4800 |

| | | | |
|---|---|---|---|
| 53 | 38 | 2 | 38 |
| | x 50 | 5 = | x 100 |
| | 1900 | 1 | 2 / 3800 = 1900 |

| | | | |
|---|---|---|---|
| 54 | 129 | 3 | 129 |
| | x 50 | 5 = | x 100 |
| | 6450 | 6 | 2 / 12,900 = 6450 |

| | | | |
|---|---|---|---|
| 55 | 432 | 0 | 432 |
| | x 50 | 5 = | x 100 |
| | 21600 | 0 | 2 / 43200 = 21600 |

| | | | |
|---|---|---|---|
| 56 | 912 | 3 | 912 |
| | x 50 | 5 = | x 100 |
| | 45600 | 6 | 2 / 91200 = 45600 |

| | | | |
|---|---|---|---|
| 57 | 1063 | 1 | 1063 |
| | x 50 | 5 = | x 100 |
| | 53,150 | 5 | 2 / 106,300 = 53,150 |

| | | | |
|---|---|---|---|
| 58 | 53 | 8 | (53 x 10) + (53 x 7) |
| | x .17 | 8 = | 530 + 371 |
| | 9.01 | 1 | 901 |

59 
| | | |
|---|---|---|
| 10.24 | 7 | (1000 x 24) + (20 x 24) + (4 x 24) |
| x .24 | 6 = | 24,000 + 480 + 96 |
| 2.4576 | 6 | 24,576 |

60 
| | | |
|---|---|---|
| 9.62 | 8 | (962 x 10) + (962 x 3) |
| x 13 | 4 = | 9620 + 2886 |
| 125.06 | 5 | 12,506 |

61 
| | | |
|---|---|---|
| 18.43 | 7 | (1800 x 8) + (40 x 8) + (3 x 8) |
| x .08 | 8 = | 14,400 + 320 + 24 |
| 1.4744 | 2 | 14,744 |

62 
| | | |
|---|---|---|
| 19.69 | 7 | (1000 x 4) + (900 x 4) + (60 x 4) + (9 x 4) |
| x .04 | 4 = | 4000 + 3600 + 240 + 36 |
| .7876 | 1 | 7876 |

63 
| | | |
|---|---|---|
| 23.16 | 3 | (2316 x 70) + (2316 x 5) |
| x .75 | 3 = | 162,120 + 11,580 |
| 17.37 | 0 | 173700 |

64 $4^3 =$ (4) (4) (4) = 64

65 $10^8 =$ (10) (10) (10) (10) (10) (10) (10) (10)
(100) (100) (100) (100)
(10,000) (10,000)
= 100,000,000

66 $9^5 =$ (9) (9) (9) (9) (9)
(81) (81) (9)
= 59049

67 $14^2 =$ (14) (14)
= 196

68 $7^1 =$ 7

69 $3^3 =$ (3) (3) (3)
= 27

70 $8^4 =$ (8) (8) (8) (8)
= 4096

71 $9^3 =$ (9) (9) (9)
= 729

72 $2.05 \times 10^3 = 2.05 \times (10)(10)(10) = 205 \times 10 = 2050$

73 $1.179 \times 10^5 = 1.179 \times (10)(10)(10)(10)(10)$
$= 1179 \times (10)(10)$
$= 117{,}900$

74 $3.201 \times 10^{-4} = 3.201 \times .0001 = .0003201$

75 $1.116 \times 10^{-8} = 1.116 \times .00000001 = .000000000116$

76 $3.14 \times 10^1 = 3.14 \times 10 = 31.4$

77 $4.20 \times 10^0 = 4.20 \times 1 = 4.20$

78 $117 \times 10^{-2} = 117 \times .01 = 1.17$

79 $401 \times 10^{-3} = 401 \times .001 = .401$

80 $14 = (7)(2)$

81 $15 = (5)(3)$

82 $27 = (3)(3)(3)$

83 $84 = (2)(42) = (2)(2)(21) = (2)(2)(3)(7)$

84 $87 = (3)(29)$

85 $90 = (2)(45) = (2)(9)(5) = (2)(3)(3)(5)$

86 $162 = (2)(81) = (2)(9)(9) = (2)(3)(3)(3)(3)$

87 $203 = (7)(29)$

[88] 445 = (5) (89)

[89] 916 = (2) (458) = (2) (2) (229)

[90] 222 = (2) (111) = (2) (3) (37)

[91] 360 = (3) (120) = (3) (3) (40)
= (3) (3) (2) (20)
= (3) (3) (2) (2) (10)
= (3) (3) (2) (2) (2) (5)

# Subtraction:

[1]
$$\begin{array}{rcr} 413 & 8 & 413 \\ \underline{-107} & 8 = & \underline{-100} \\ 306 & 0 & 313 \\ & & \underline{-7} \\ & & 306 \end{array}$$

[2]
$$\begin{array}{rc} 227 & 2 \\ \underline{-113} & 5 \\ 114 & 6 \end{array}$$

[3]
$$\begin{array}{rcr} 451 & 1 & 451 \\ \underline{-22} & 4 = & \underline{-20} \\ 429 & 6 & 431 \\ & & \underline{-2} \\ & & 429 \end{array}$$

[4]
$$\begin{array}{rc} 508 & 4 \\ \underline{-38} & 2 \\ 470 & 2 \end{array}$$

[5]
$$\begin{array}{rcr} 612 & 0 & 612 \\ \underline{-147} & 3 = & \underline{-150} \\ 465 & 6 & 462 \\ & & \underline{+3} \\ & & 465 \end{array}$$

[6]
$$\begin{array}{rcr} 901 & 1 & 900 \\ \underline{-399} & 3 = & \underline{-400} \\ 502 & 7 & 500 \\ & & \underline{+2} \\ & & 502 \end{array}$$

[7]
$$\begin{array}{rcr} 612 & 0 & 612 \\ \underline{-39} & 3 = & \underline{-40} \\ 573 & 6 & 572 \\ & & \underline{+1} \\ & & 573 \end{array}$$

[8]
$$\begin{array}{rcr} 451 & 1 & 451 \\ \underline{-112} & 4 = & \underline{-110} \\ 339 & 6 & 341 \\ & & \underline{-2} \\ & & 339 \end{array}$$

**9**
$$\begin{array}{rc} 637 & 7 \\ -201 & 3 \\ \cline{1-1} 436 & 4 \end{array}$$

**10**
$$\begin{array}{rcr} 891 & 0 & 891 \\ -14 & 5 = & -10 \\ \cline{1-1}\cline{3-3} 877 & 4 & 881 \\ & & -4 \\ \cline{3-3} & & 877 \end{array}$$

**11**
$$\begin{array}{rcr} 917 & 8 & 917 \\ -38 & 2 = & -50 \\ \cline{1-1}\cline{3-3} 879 & 6 & 867 \\ & & +12 \\ \cline{3-3} & & 879 \end{array}$$

**12**
$$\begin{array}{rcr} 723 & 3 & 727 \\ -27 & 0 = & -27 \\ \cline{1-1}\cline{3-3} 696 & 3 & 700 \\ & & -4 \\ \cline{3-3} & & 696 \end{array}$$

**13**
$$\begin{array}{rcr} 119 & & 120 \\ -201 & = & -200 \\ \cline{1-1}\cline{3-3} -82 & & -80 \\ & & -2 \\ \cline{3-3} & & -82 \end{array}$$

**14**
$$\begin{array}{rcr} 338 & & 338 \\ -600 & = & -638 \\ \cline{1-1}\cline{3-3} -262 & & -300 \\ & & +38 \\ \cline{3-3} & & -262 \end{array}$$

**15**
$$\begin{array}{rcr} 391 & & 400 \\ -712 & = & -712 \\ \cline{1-1}\cline{3-3} -321 & & -312 \\ & & -9 \\ \cline{3-3} & & -321 \end{array}$$

**16**
$$\begin{array}{rcr} 493 & & 500 \\ -508 & = & -508 \\ \cline{1-1}\cline{3-3} -15 & & -8 \\ & & -7 \\ \cline{3-3} & & -15 \end{array}$$

**17**
$$\begin{array}{rcr} 327 & & 328 \\ -618 & = & -618 \\ \cline{1-1}\cline{3-3} -291 & & -290 \\ & & -1 \\ \cline{3-3} & & -291 \end{array}$$

**18**
$$\begin{array}{rcr} 112 & & 100 \\ -300 & = & -300 \\ \cline{1-1}\cline{3-3} -188 & & -200 \\ & & +12 \\ \cline{3-3} & & -188 \end{array}$$

**19**
$$\begin{array}{rc} 429 & 6 \\ -113 & 5 \\ \cline{1-1} 316 & 1 \end{array}$$

**20**
$$\begin{array}{rc} 398 & 2 \\ -107 & 8 \\ \cline{1-1} 291 & 3 \end{array}$$

**21**
$$\begin{array}{rcr} 612 & 0 & 607 \\ -457 & 7 = & -457 \\ \cline{1-1}\cline{3-3} 155 & 2 & 150 \\ & & +5 \\ \cline{3-3} & & 155 \end{array}$$

**22**
$$\begin{array}{rc} 107 & 8 \\ -43 & 7 \\ \cline{1-1} 64 & 1 \end{array}$$

| 23 | 711 | 0 | 717 |
|---|---|---|---|
| | – 17 | 8 = | – 17 |
| | 694 | 1 | 700 |
| | | | – 6 |
| | | | 694 |

| 24 | 1031 | 5 |
|---|---|---|
| | – 227 | 2 |
| | 804 | 3 |

| 25 | 1236 | 3 |
|---|---|---|
| | – 908 | 8 |
| | 328 | 4 |

| 26 | 723 | 3 | 723 |
|---|---|---|---|
| | – 92 | 2 = | – 100 |
| | 631 | 1 | 623 |
| | | | + 8 |
| | | | 631 |

| 27 | 1630 | 1 |
|---|---|---|
| | – (–29) | 2 |
| | 1659 | 3 |

| 28 | 48 | 3 |
|---|---|---|
| | – (–50) | 5 |
| | 98 | 8 |

| 29 | 56 | 2 |
|---|---|---|
| | – (–93) | 3 |
| | 149 | 5 |

| 30 | 18 | 0 |
|---|---|---|
| | – (–32) | 5 |
| | 50 | 5 |

| 31 | 64 | 1 |
|---|---|---|
| | + (–12) | 3 |
| | 52 | 7 |

| 32 | 169 | 7 |
|---|---|---|
| | + (–12) | 3 |
| | 157 | 4 |

| 33 | 383 | 5 |
|---|---|---|
| | + (–72) | 0 |
| | 311 | 5 |

| 34 | 99 | 0 |
|---|---|---|
| | + (–18) | 0 |
| | 81 | 0 |

| 35 | 36 | 0 |
|---|---|---|
| | x (–12) | 3 |
| | – 432 | 0 |

| 36 | –18 | 0 | –10 | | –8 |
|---|---|---|---|---|---|
| | x 13 | 4 = | 13 | + | x 13 |
| | –234 | 0 | (–130) | + | (–104) |

| 37 | –12 | 3 | –6 | | |
|---|---|---|---|---|---|
| | x –16 | 7 = | x –16 | | |
| | 192 | 3 | 2 (96) | = | 192 |

38

$$\begin{array}{r} -36 \\ \underline{x\,-7} \\ 252 \end{array} \quad \begin{array}{l} 0 \\ 7 = \\ 0 \end{array} \quad \begin{array}{r} -30 \\ \underline{x\,-7} \\ 210 \end{array} \quad \begin{array}{c} \\ + \\ + \end{array} \quad \begin{array}{r} -6 \\ \underline{x\,-7} \\ 42 \end{array}$$

39

$$\begin{array}{r} 98.01 \\ \underline{-\,45.37} \\ 52.64 \end{array} \quad \begin{array}{l} 0 \\ 1 = \\ 8 \end{array} \quad \begin{array}{r} 98.37 \\ -\,\underline{45.37} \\ 53.00 \\ \underline{-\,.36} \\ 52.64 \end{array}$$

40

$$\begin{array}{r} 36.98 \\ \underline{-\,12.73} \\ 24.25 \end{array} \quad \begin{array}{l} 8 \\ 4 \\ 4 \end{array}$$

41

$$\begin{array}{r} 37.96 \\ \underline{-\,4.12} \\ 33.84 \end{array} \quad \begin{array}{l} 7 \\ 7 \\ 0 \end{array}$$

42

$$\begin{array}{r} 12.18 \\ \underline{-\,16.42} \\ -\,4.24 \end{array} \quad \begin{array}{l} 3 \\ 4 = \\ 1 \end{array} \quad \begin{array}{r} 12.42 \\ \underline{-\,16.42} \\ -\,4.00 \\ \underline{-\,.24} \\ -\,4.24 \end{array}$$

43

$$\begin{array}{r} 112.39 \\ \underline{-\,48.42} \\ 63.97 \end{array} \quad \begin{array}{l} 7 \\ 0 \\ 7 \end{array}$$

44

$$\begin{array}{r} 169.53 \\ \underline{-\,32.47} \\ 137.06 \end{array} \quad \begin{array}{l} 6 \\ 7 \\ 8 \end{array}$$

45

$$\begin{array}{r} 403.89 \\ \underline{-\,107.12} \\ 296.77 \end{array} \quad \begin{array}{l} 6 \\ 2 \\ 4 \end{array}$$

46

$$\begin{array}{r} 223.12 \\ \underline{-\,162.41} \\ 60.71 \end{array} \quad \begin{array}{l} 1 \\ 5 \\ 5 \end{array}$$

# Division:

1 $$\begin{array}{r} 4 \\ 7\overline{)\,28} \end{array}$$

2 $$\begin{array}{r} 9 \\ 6\overline{)\,54} \end{array}$$

3 $$\begin{array}{r} 15 \\ 5\overline{)\,75} \\ \underline{-5} \\ 25 \\ \underline{-25} \\ 0 \end{array}$$

4 $$\begin{array}{r} 12 \\ 9\overline{)\,108} \\ \underline{-9} \\ 18 \\ \underline{-18} \\ 0 \end{array}$$

5 $$\begin{array}{r} 23 \\ 10\overline{)\,230} \\ \underline{-20} \\ 30 \\ \underline{-30} \\ 0 \end{array}$$

6 $$\begin{array}{r} 5 \\ 18\overline{)\,90} \\ \underline{-90} \\ 0 \end{array}$$

7 $$\begin{array}{r} 1.64 \\ 14\overline{)\,23.0} \\ \underline{-14} \\ 90 \\ \underline{-84} \\ 60 \\ \underline{-56} \\ 4 \end{array}$$

8 $$\begin{array}{r} .66 \\ 15\overline{)\,10.00} \\ \underline{-90} \\ 100 \\ \underline{-90} \\ 10 \end{array}$$

9 $$\begin{array}{r} 11.5 \\ 8\overline{)\,92.0} \\ \underline{-8} \\ 12 \\ \underline{-8} \\ 40 \\ \underline{-40} \\ 0 \end{array}$$

10 $$\begin{array}{r} .025 \\ 400\overline{)\,10.00} \\ \underline{-800} \\ 2000 \\ \underline{-2000} \\ 0 \end{array}$$

11 $$\begin{array}{r} 1 \\ 25\overline{)\,25} \\ \underline{-25} \\ 0 \end{array}$$

12 $\frac{1}{4} = .25$

$$\begin{array}{r} .25 \\ 4\overline{)\,1.00} \\ \underline{-8} \\ 20 \\ \underline{-20} \\ 0 \end{array}$$

13 $\frac{3}{8} = .375$

$$\begin{array}{r} .375 \\ 8\overline{)\,3.00} \\ \underline{-24} \\ 60 \\ \underline{-56} \\ 40 \\ \underline{-40} \\ 0 \end{array}$$

[14] $\frac{6}{12} = .5$

$$\begin{array}{r} .5 \\ 12 \overline{)\ 6.0} \\ -6\,0 \\ \hline 0 \end{array}$$

[15] $\frac{4}{9} = .44$

$$\begin{array}{r} .44 \\ 9 \overline{)\ 4.00} \\ -3\,60 \\ \hline 40 \\ -40 \\ \hline 0 \end{array}$$

[16] $\frac{3}{2} = 1.5$

$$\begin{array}{r} 1.5 \\ 2 \overline{)\ 3.0} \\ -2 \\ \hline 10 \\ -10 \\ \hline 0 \end{array}$$

[17] $\frac{8}{12} = .66$

$$\begin{array}{r} .66 \\ 12 \overline{)\ 8.0} \\ -7\,2 \\ \hline 80 \\ -72 \\ \hline 6 \end{array}$$

[18] $\frac{7}{8} = .875$

$$\begin{array}{r} .875 \\ 8 \overline{)\ 7.000} \\ -64 \\ \hline 60 \\ -56 \\ \hline 40 \\ -40 \\ \hline 0 \end{array}$$

[19] $\frac{1}{3} = .33$

$$\begin{array}{r} .33 \\ 3 \overline{)\ 1.00} \\ -9 \\ \hline 10 \\ -9 \\ \hline 1 \end{array}$$

[20] $\frac{4}{3} = 1.33$

$$\begin{array}{r} 1.33 \\ 3 \overline{)\ 4.00} \\ -3 \\ \hline 10 \\ -9 \\ \hline 10 \\ -9 \\ \hline 1 \end{array}$$

[21] $\frac{12}{8} = 1.5$

$$\begin{array}{r} 1.5 \\ 8 \overline{)\ 12.0} \\ -8 \\ \hline 40 \\ -40 \\ \hline 0 \end{array}$$

[22] $\frac{9}{6} = 1.5$

$$\begin{array}{r} 1.5 \\ 6 \overline{)\ 9.0} \\ -6 \\ \hline 3\,0 \\ -3\,0 \\ \hline 0 \end{array}$$

[23] $\frac{7}{5} = 1.4$

$$\begin{array}{r} 1.4 \\ 5 \overline{)\ 7.0} \\ -5 \\ \hline 2\,0 \\ -2\,0 \\ \hline 0 \end{array}$$

24 $5\frac{1}{2} = \frac{10}{2} + 1/2 = \frac{11}{2} = 5.5$

$$\begin{array}{r} 5.5 \\ 2\,\overline{)\,11.0} \\ \underline{-10} \\ 10 \\ \underline{-10} \\ 0 \end{array}$$

25 $6\frac{1}{3} = \frac{18}{3} + \frac{1}{3} = \frac{19}{3} = 6.33$

$$\begin{array}{r} 6.33 \\ 3\,\overline{)\,19.00} \\ \underline{-18} \\ 10 \\ \underline{-9} \\ 10 \\ \underline{-9} \\ 1 \end{array}$$

26 $2\frac{3}{4} = \frac{8}{4} + \frac{3}{4} = \frac{11}{4} = 2.75$

$$\begin{array}{r} 2.75 \\ 4\,\overline{)\,11.00} \\ \underline{-8} \\ 30 \\ \underline{-28} \\ 20 \\ \underline{-20} \\ 0 \end{array}$$

27 $3 + \frac{5}{7} = 21/7 + \frac{5}{7} = \frac{26}{7} = 3.71$

$$\begin{array}{r} 3.71 \\ 7\,\overline{)\,26.00} \\ \underline{-21} \\ 50 \\ \underline{-49} \\ 10 \\ \underline{-7} \\ 3 \end{array}$$

28 $\sqrt{9} = 3$

29 $\sqrt{16} = 4$

30 $\sqrt{25} = 5$

31 $\sqrt{36} = 6$

[32] $\sqrt{46} = 6.83$

$$\begin{array}{r} 7.66 \\ 6\,\overline{)\,46.00} \\ \underline{-42\phantom{.00}} \\ 40\phantom{0} \\ \underline{-36\phantom{0}} \\ 40 \\ \underline{-36} \\ 4 \end{array}$$

$$\begin{array}{r} 7.66 \\ +\,6.00 \end{array}$$

$2\,/\,13.66 = 6.83$

[33] $\sqrt{84} = 9.165$

$$\begin{array}{r} 9.33 \\ 9\,\overline{)\,84.00} \\ \underline{-81\phantom{.00}} \\ 30\phantom{0} \\ \underline{-27\phantom{0}} \\ 30 \\ \underline{-27} \\ 3 \end{array}$$

$$\begin{array}{r} 9.33 \\ +\,9.00 \end{array}$$

$2\,/\,18.33 = 9.165$

[34] $\sqrt{104} = 10.2$

$$\begin{array}{r} 10.4 \\ 10\,\overline{)\,104.0} \\ \underline{-10\phantom{4.0}} \\ 04\phantom{0} \\ \underline{-0\phantom{0}} \\ 40 \\ \underline{-40} \\ 0 \end{array}$$

$$\begin{array}{r} 10.40 \\ +\,10.00 \end{array}$$

$2\,/\,20.40 = 10.2$

[35] $\sqrt{63} = 7.9$

$$\begin{array}{r} 9 \\ 7\,\overline{)\,63} \end{array}$$

$$\begin{array}{r} 9.00 \\ +\,7.00 \end{array}$$

$2\,/\,16.00 = 8$

[36] $\frac{4}{5} + \frac{7}{5} = \frac{11}{5} = 2\frac{1}{5}$

[37] $\frac{5}{4} + \frac{3}{2} = \frac{5}{4} + \frac{6}{4} = \frac{11}{4} = 2\frac{3}{4}$

[38] $\frac{3}{4} + \frac{1}{3} = \frac{9}{12} + \frac{4}{12} = \frac{13}{12} = 1\frac{1}{12}$

[39] $\frac{12}{7} + \frac{2}{5} = \frac{60}{35} + \frac{14}{35} = \frac{74}{35} = 2\frac{4}{35}$

40. $\frac{9}{10} + \frac{7}{8} = \frac{72}{80} + \frac{70}{80} = \frac{142}{80} = 1\frac{62}{80} = 1\frac{31}{40}$

41. $\frac{11}{8} + \frac{8}{11} = \frac{121}{88} + \frac{64}{88} = \frac{185}{88} = 2\frac{9}{88}$

42. $10\frac{1}{3} + \frac{6}{5} = \frac{31}{3} + \frac{6}{5} = \frac{155}{15} + \frac{18}{15} = \frac{173}{15} = 11\frac{8}{15}$

43. $\frac{3}{4} + \frac{1}{8} = \frac{6}{8} + \frac{1}{8} = \frac{7}{8}$

44. $7\frac{5}{6} + 3\frac{1}{2} = \frac{47}{6} + \frac{7}{2} = \frac{47}{6} + \frac{21}{6} = \frac{68}{6} = 11\frac{2}{6} = 11\frac{1}{3}$

45. $4\frac{7}{6} + 5\frac{1}{6} = \frac{31}{6} + \frac{31}{6} = \frac{62}{6} = 10\frac{2}{6} = 10\frac{1}{3}$

46. $\frac{6}{8} \times \frac{3}{4} = \frac{18}{32} = \frac{9}{16}$

47. $\frac{4}{6} \times \frac{1}{3} = \frac{4}{18} = \frac{2}{9}$

48. $\frac{5}{6} \times \frac{2}{3} = \frac{10}{18} = \frac{5}{9}$

49. $\frac{1}{2} \times \frac{5}{8} = \frac{5}{16}$

50. $\frac{1}{4} \times \frac{1}{8} = \frac{1}{32}$

51. $\frac{1}{6} \times \frac{2}{3} = \frac{2}{18} = \frac{1}{9}$

52. $\frac{7}{8} \times \frac{11}{12} = \frac{77}{96}$

53. $\frac{6}{7} \times \frac{3}{4} = \frac{18}{28} = \frac{9}{14}$

54. $\frac{10}{7} \times \frac{4}{5} = \frac{40}{35} = \frac{8}{7} = 1\frac{1}{7}$

55. $\frac{11}{7} \times \frac{5}{6} = 1\frac{13}{42}$

56. $\frac{6}{8} - \frac{3}{4} = \frac{6}{8} - \frac{6}{8} = 0$

57. $\frac{6}{7} - \frac{3}{4} = \frac{24}{28} - \frac{21}{28} = \frac{3}{28}$

58. $\frac{7}{8} - \frac{11}{12} = \frac{84}{96} - \frac{88}{96} = -\frac{4}{96} = -\frac{1}{24}$

59. $\frac{9}{4} - \frac{1}{3} = \frac{27}{12} - \frac{4}{12} = \frac{23}{12} = 1\frac{11}{12}$

60 $\frac{10}{7} - \frac{4}{5} = \frac{50}{35} - \frac{28}{35} = \frac{22}{35}$

61 $\frac{7}{8} - \frac{2}{6} = \frac{21}{24} - \frac{8}{24} = \frac{13}{24}$

62 $4\frac{1}{5} - 3\frac{1}{3} = \frac{21}{5} - \frac{10}{3} = \frac{63}{15} - \frac{50}{15} = \frac{13}{15}$

63 $5\frac{1}{4} - \frac{3}{5} = \frac{21}{4} - \frac{3}{5} = \frac{105}{20} - \frac{12}{20} = \frac{93}{20} = 4\frac{13}{20}$

64 $3\frac{7}{8} - 2\frac{5}{9} = \frac{31}{8} - \frac{23}{9} = \frac{279}{72} - \frac{184}{72} = \frac{95}{72} = 1\frac{23}{72}$

65 $3\frac{3}{7} - 2\frac{2}{3} = \frac{24}{7} + \frac{8}{3} = \frac{72}{21} - \frac{56}{21} = \frac{16}{21}$

66 $\frac{4}{3} \div \frac{2}{3} = \frac{12}{6} = 2$

67 $\frac{1}{3} \div \frac{1}{2} = \frac{2}{3}$

68 $\frac{4}{8} \div \frac{1}{5} = \frac{20}{8} = 2\frac{4}{8} = 2\frac{1}{2}$

69 $\frac{5}{7} \div \frac{6}{2} = \frac{10}{42} = \frac{5}{21}$

70 $\frac{9}{11} \div \frac{1}{7} = \frac{63}{11}$

71 $\frac{3}{8} \div \frac{5}{6} = \frac{18}{40} = \frac{9}{20}$

72 $\frac{7}{9} \div \frac{1}{3} = \frac{21}{9} = 2\frac{3}{9} = 2\frac{1}{3}$

73 $\frac{6}{7} \div \frac{1}{4} = \frac{24}{7} = 3\frac{3}{7}$

74 $-312 \div (-4) = 78$

75 $-69 \div (-3) = 23$

76 $-49 \div (-7) = 7$

77 $306 \div (-12) = -25.5$

78 $-465 \div (15) = -31$

79 $510 \div (-10) = -51$

80 $-35 \div (-5) = 7$

81 $-175 \div (25) = -7$

82 $450 \div (-75) = -6$

83 $-759 \div (33) = -23$

84 $-96 \div (-6) = 16$

85 $156 \div (-4) = -39$

# Putting It All Together:

1.
$$\begin{array}{r} 9 \\ -24 \\ \hline -15 \end{array}$$

2.
$$\begin{array}{r} 8 \\ -48 \\ \hline -40 \end{array}$$

3.
$$\begin{array}{r} 41 \\ \times 31 \\ \hline 1271 \end{array} \quad \begin{array}{r} 5 \\ 4 = \\ 2 \end{array} \quad \begin{array}{r} 40 \\ \times 31 \\ \hline 1240 \end{array} \quad + \quad \begin{array}{r} 1 \\ \times 31 \\ \hline 31 \end{array}$$

4. $5 \times 2 \times 3 = 30$

5. $(144 + 9) - (64 + 4) =$
$$\begin{array}{r} 153 \\ -68 \\ \hline 85 \end{array} \quad \begin{array}{r} 0 \\ 5 \\ 4 \end{array} \quad \begin{array}{r} 158 \\ -68 \\ \hline 90 \\ -5 \\ \hline 85 \end{array}$$

6. $(81 - 9) \div (3 + 25) =$
$$28 \overline{)\,72.} = 2.57$$
$$\begin{array}{r} 72. \\ -56 \\ \hline 160 \\ -140 \\ \hline 200 \\ -196 \\ \hline \end{array}$$

7. $6^2 + 96 =$
$$\begin{array}{r} 36 \\ +96 \\ \hline 132 \end{array} \quad \begin{array}{r} 0 \\ 6 \\ 6 \end{array}$$

8. $5^2 + 15 =$
$$\begin{array}{r} 25 \\ +15 \\ \hline 40 \end{array} \quad \begin{array}{r} 7 \\ 6 \\ 4 \end{array}$$

9. $\frac{(24)(4)}{2} = \frac{96}{2} = 48$

10. $\frac{300}{30} = 10$

11. $\frac{4^2 + 13}{6} = \frac{29}{6} = 4.83$

12. $\frac{4^3 + (-16)}{8} = \frac{48}{8} = 6$

13. $16 + 4 + \frac{1}{4} = 20\frac{1}{4}$

14. $8 + 64 + \frac{2}{4} = 72\frac{1}{2}$

15. $148 + 4 + 4 = 156$

16. $60 + 32 - 10 = 82$

17. $64 + 48 - 4 = 108$

18. $64 - 32 + 4 = 36$

19. $\frac{(18)(52)}{3} = \frac{936}{3} = 312$

20. $\frac{(14)(18)}{12} = \frac{252}{12} = 21$

21. $\frac{168 + 12 - (18)}{4} = \frac{162}{4} = 40\frac{1}{2}$

22. $\frac{(80 - 12) \div (10)}{4} = 1.7$

23. $(5 + 7)(5 - 7) = 12 \times (-2) = -24$

24. $10 + 35 - 21 = 24$

25. $\frac{(25 + 7) - 343}{7} = \frac{-311}{7} = -44.42$

26. $\frac{(25 + 49)(25 - 49)}{25} = \frac{74 \times (-24)}{25} = \frac{-1776}{25} = -71.04$

27. $\frac{375 + 14}{21} = \frac{389}{21} = 18.52$

28. $\frac{50 - 105}{10} = \frac{-55}{10} = -5.5$

29. $169 + \frac{9}{49} = 169\frac{9}{49}$

30. $169 - \frac{21}{7} = 169 - 3 = 166$

31. $507 + \frac{18}{49} = 507\frac{18}{49}$

32. $\frac{169}{6} + \frac{27}{49} = \frac{8443}{294} = 28.72$

33. $(\frac{39}{7})(\frac{3}{14}) = \frac{117}{98} = 1.19$

34. $9 \div \frac{45}{49} = \frac{9}{1} \div \frac{45}{49} = \frac{441}{45} = 9.8$

[35] $3x^2 = -13$
$x^2 = -13/3$
$x = \sqrt{-13/3}$

[36] $x^2 - 2x = 0$
$x^2 = 2x$
$x = 2$

[37] $12x = -4$
$3x = -1$
$x = -1/3$

[38] $4x = 12$
$x = 3$

[39] $10x \div 5 = 0$
$10x = 0$
$x = 0$

[40] $3x^2 = 12$
$x^2 = 4$
$x = 2$

[41] $x\left(\frac{5x^2 + 3x}{x}\right) = 0(x)$

$5x^2 + 3x = 0$
$5x^2 = -3x$
$5x = -3$
$x = -3/5$

[42] $\left(\frac{2x^2 - x}{4}\right)4 = 0(4)$

$2x^2 - x = 0$
$2x^2 = x$
$2x = 1$
$x = 1/2$

# Story Problems:

[1]

$$432 \text{ Dollars} = 432 \text{ Dollars} \times \frac{1 \text{ Thud}}{27 \text{ Dollars}} = \frac{432 \text{ Thuds}}{27} = 16 \text{ Thuds}$$

[2]

$$432 \text{ Dollars} = 432 \text{ Dollars} \times \frac{3 \text{ Thuds}}{75 \text{ Dollars}} = \frac{1296 \text{ Thuds}}{75} = 17.28 \text{ Thuds}$$

17.28 Thuds – 16 Thuds = 1.28 extra

[3]

$\frac{10 \text{ miles}}{30 \text{ mph}}$ = 1/3 hour, or 20 minutes

$\frac{10 \text{ miles}}{20 \text{ mph}}$ = 1/2 hour, or 30 minutes

[4]

30 x 50 lbs. = 1500 lbs.

$$1500 \text{ lbs.} \times \frac{1 \text{ kg}}{2.2} = \frac{1500 \text{ kg}}{2.2} = 681.82 \text{ kg}$$

5

$$300 \text{ cucumbers} \times \frac{1 \text{ bushel corn}}{30 \text{ cucmbers}} = \frac{300 \text{ bushels}}{30} = 10 \text{ bushels}$$

$$300 \text{ cucumbers} \times \frac{1 \text{ bushel corn}}{30 \text{ cucumbers}} \times \frac{12 \text{ lbs. honey}}{1 \text{ bushel corn}} =$$

$$\frac{3600 \text{ lbs. honey}}{30} = 120 \text{ lbs. honey}$$

1 side of beef = 10 bushels corn, so according to the first part of the puzzle, yes.

$$\frac{2 \text{ sides beef}}{1 \text{ oz. gold}} \times \frac{2 \text{ oz. gold}}{1 \text{ horse}} = \frac{4 \text{ sides beef}}{1 \text{ horse}} \text{ ( or, 4)}$$

$$\frac{4 \text{ sides beef}}{1 \text{ horse}} \times \frac{10 \text{ bushels corn}}{1 \text{ side beef}} \times \frac{30 \text{ cucumbers}}{1 \text{ bushel corn}} = \frac{1200 \text{ cucumbers}}{1 \text{ horse}}$$

6

$T = Prt$
$T = 3500(.039)(.5)$
$T = 3500(.0195) = \$68.25$

$T = Prt$
$T = 3500(.044)(1)$
$T = \$154$

7

$T = Prt$
$T = 3500(.06)(.583)$
$T = \$122.43$, or not more than the 12 month CD

8

$$\frac{186{,}000 \text{ miles}}{1 \text{ second}} \times \frac{60 \text{ seconds}}{1 \text{ minute}} \times \frac{60 \text{ minutes}}{1 \text{ hour}} \times \frac{24 \text{ hours}}{1 \text{ day}} \times \frac{7 \text{ days}}{1 \text{ week}}$$

(1.86 x 105)(60)(60)(24)(7) = 1.125 x 1011 miles

9

$$\frac{1{,}000{,}000{,}000 \text{ km}^3}{1 \text{ ocean}} \times \frac{1{,}000{,}000 \text{ cm}^3}{1 \text{ km}^3} \times \frac{1 \text{ Pepsi can}}{355 \text{ cm}^3}$$

$$\frac{(1 \times 10^9)(1 \times 10^6)(1)}{355} = \frac{1 \times 10^{15}}{355} = \text{approximately } 2.82 \times 10^{12} \text{ cans}$$

# Math in Real Life:

1.

| | | | |
|---|---|---|---|
| coffee | $ 4.75 | $20.00 | subtotal |
| eggs | 1.25 | x .06 | tax |
| milk | 3.25 | 1.8000 | |
| bread | 1.25 | | |
| newspaper | .50 | $20.00 | |
| dogfood | 9.00 | + 1.80 | |
| subtotal | $20.00 | $21.80 | |

You are short by $1.80

2. 16 oz. jar = $.11 per ounce  28 oz. jar = $.08 per ounce

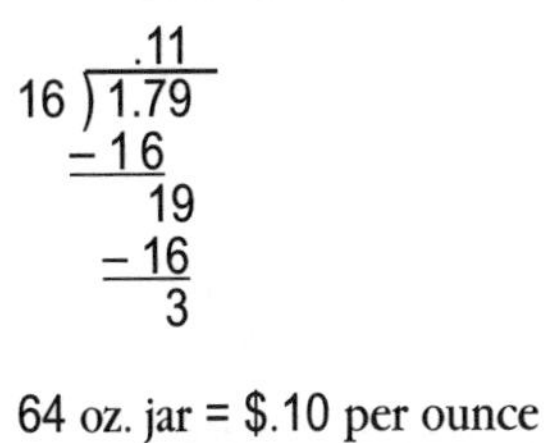

```
      .08
28 )2.29
   -2 24
       5
```

64 oz. jar = $.10 per ounce  28 oz. is best deal

```
      .102
64 )6.590
   -6 4
      190
     -128
       62
```

3. Estimated total diner bill = $ 52.00
(6%) estimated tax + 3.00
total dinner bill = $ 55.00

Actual prices:

| | |
|---|---|
| | $ 12.95 |
| | 1.25 |
| | 2.25 |
| | 6.95 |
| | .85 |
| | 13.95 |
| | 1.15 |
| | 2.25 |
| | 4.95 |
| | 3.95 |
| | 1.25 |
| Total Dinner Bill | $ 51.75 |
| Tax 6% | 3.10 |
| | $ 54.85 |

4. 10 min. at 35 mph — $\frac{1}{6}$ hour x $\frac{35 \text{ miles}}{\text{hours}} = \frac{35}{6}$ = 5.8 miles

20 min at 80 mph — $\frac{2}{6}$ hour x $\frac{80 \text{ miles}}{\text{hour}} = \frac{160}{6}$ = 26.6 miles

10 min stopped — $\frac{1}{6}$ hour x $\frac{0 \text{ miles}}{\text{hour}} = \frac{0}{6}$ = 0 miles

20 min at 55 mph — $\frac{2}{6}$ hour x $\frac{55 \text{ miles}}{\text{hour}} = \frac{110}{6}$ = 18.3 miles

60 min = 1 hour — Bruce traveled 50.7 miles

5.
70 feet per side
x 4 # of sides
280
x 3 rows of boards
840

Farmer Frank needs 840 feet of boards.

6.
840 (total footage)
(840 x .30) + (840 x .05)
x .35 (per foot)
$294.00 cost of boards

= 252.00 + 42.00
294.00

7.
70
– 1
69
x 4
276 feet of wire

8.
276 feet
x .03 per foot
$8.19 cost for wire

$ 8.19 for wire
$294.00 for boards
$302.19 total cost of fence

9.
$3000.00
x .055
$165.00 interest

10.
$3000.00
x .068
$204.00 interest for 1 year
x 5
$1020.00 interest for 5 years

[11] $\pi r^2$ = area of a circle
$2 \pi r$ = perimeter
r = .5 distance across (diameter)

3.14 x 32 = 3.14 x 9 = 28.26 square inches in area

2 x 3.14 x 3 = 3.14 x 6 = 6.28 inches in perimeter

[12] 1.86 x 105 x # of seconds in a year:

| | |
|---|---|
| 60 seconds | (1 hour) |
| x 24 | (hours/day) |
| 1440 | |
| x 365 | (days/year) |
| 525,600 | (seconds/year) |

1.86 x 105 x 525,600 = 977,616 x 105
= 977,616,000,000 light years

# Math and your bank:

[1] $T = P(1 + I)^n$
$T = 5000(1 + .0047)^{12}$
T = 5287.30, or $287.30 in interest

$T = P(1 + I)^n$
$T = 5000(1 + .01425)^4$
T = 5291.15, or $291.15 in interest

The 2nd CD is the better deal.

[2] $$T = (150{,}000)\left(\frac{I(1 + I)^n}{(1 + I)^n - 1}\right)(360)$$

$$T = (150{,}000)\left(\frac{.00833\,(1.00833)^{360}}{(1.00833)^{360} - 1}\right)(360)$$

$$T = (150{,}000)\left(\frac{.165049}{18.813805}\right)(360)$$

T = (150,000) (.008773) (360) = $473,742.00

$$T = (150{,}000) \left( \frac{.0075\ (1.0075)^{360}}{(1.0075)^{360} - 1} \right) (360)$$

$$T = (150{,}000) \left( \frac{.1104793}{13.730576} \right) (360) = \$434{,}496.13$$

$473,742.00 – $434,496.13 = $39,245.87 saved

1000 (1 + I)k – 1000 = interest
1000 (1.015)12 – 1000 = $195.62 interest / year, or 19.6%

4 20.00 – 18.25 = % increase = .096, or
9.6% = 18.25

1000 (1 + .01125)4 – 1000 = $45.76 interest, or 4.58%

The stock is over twice as good.

5 Payment = (270,000) (.0091666)(1.0091666)360
(1.0091666)360 –1
= (270,000) .2448224 = 24.707462

Payment = $2,675.39 per month for 30 years

Payment = (270,000) (.0091666)(1.0091666)240
= (1.0091666)240 –1
= (270,000) .0819024
7.9348737

Payment = $2,786.89 per month for 20 years
Payment = (270,000) (.0091666)(1.0091666)120
(1.0091666)120 –1
= (270,000) .0274001
1.9891259
Payment = $3,719.24 per month for 10 years

6 T = Prt
T = (25,000)(.09)(4)
T = $9,000 interest paid
T = P (1 + I)$^n$ – P
T =(25,000)(1.0075)48 – P

T = 35,785.13 – 25,000 = $10,785.13 interest paid

He would save $1,785.13

$T = P(1 + I)^n - P$
T = 500 (1 + .0145833)12 – P
T = 594.87 – 500 = $94.87 saved per year

# Resources

## ••• Recommended Reading •••

### *Accelerative Learning/Intelligence*

Scheele, Paul. *Natural Brilliance*. Minneapolis, MN: Learning Strategies Corporation, 1997.

Wenger, Win, and Poe, Richard. *The Einstein Factor*. Rocklin, CA: Prima Publishing, 1996.

Witt, Scott. *How To Be Twice As Smart*. West Nyack, NY: Parker Publishing Company, Inc., 1983.

## General Mathematics

Benjamin, Arthur, and Shermer, Michael. *Mathemagics*. Los Angeles, CA: Lowell House, 1994.

Kogelman, Stanley and Heller, Barbara. *The Only Math Book You'll Ever Need*. New York, NY: HarperCollins, 1986.

## Probability And Statistics

Huff, Darrell. *How To Lie With Statistics*. New York, NY: W. W. Norton & Company, Inc., 1954.

McGervey, John. *Probabilities In Everyday Life*. New York, NY: Ivy Books, 1986.

## ••• Resources •••

For information on audio-programs and seminars for your personal and professional development, contact:

Learning Strategies Corporation
900 East Wayzata Boulevard
Wayzata, Minnesota 55391-1836 USA
952-476-9200
800-735-8273
Fax: 952-475-2373
E-mail: Info@LearningStrategies.com

***www.LearningStrategies.com***

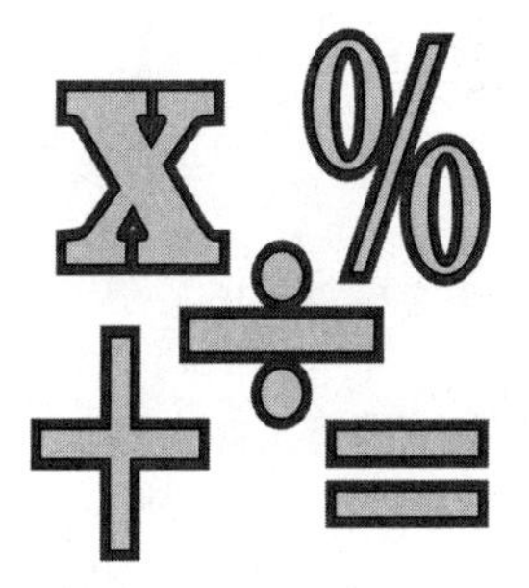

# Index

## A-M•••••



## N-W•••••